ENVIRONMENT DISASTERS AND THE LAW

Vinod Shankar Mishra
Law School,
Banaras Hindu University
Varanasi

ASHISH PUBLISHING HOUSE
8/81, PUNJABI BAGH, NEW DELHI - 110 026

Published by :

S.B. Nangia
for **Ashish Publishing House**
8/81, Punjabi Bagh
New Delhi - 110026
Tele. 50 05 81
541 09 24

Show room :
5, Ansari Road,
Darya Ganj
New Delhi - 110002
Tele. 3274050

ISBN 81-7024-629-6

2026

Printed at :
Efficient Offset Printers,
New Delhi

Dedicated to my parents

Preface

The greatest evil and worst form of environmental pollution is caused by hazardous industries. The recent liberalization of economy has opened the flood gates for establishment of hazardous industries by National as well as multinational companies. The biggest threat to human lives and environment comes from hazardous industries but unfortunately, our policy makers have completely ignored this impact of indiscriminate industrialization. This apathy of legislature and executive have led us to worst form of industrial disasters and they are equally shared by public and private sectors. They are exemplified by Panipat disaster related to ammonia leak (1992), Grasim Mill in Gwalior (1991) and Oleum leak from Sriram (1985) and the most tragic MIC from Union Carbide Plant in Bhopal (1984). However, neither the Government nor the industries seem to have learnt any lesson from these painful experiences. Although the legislature has shown its concern by introducing the National Environmental Tribunal Bill, 1992 but it has yet to become law and sooner it is passed better, it would be for the protection and preservation of human environment. In order to understand the role of hazardous industries in the field of environmental pollution, present work makes a specific study of the oleum leak from Sriram industries. The present work illustrates the ramification of indiscriminate assault on the Indian environment. The present study adopts an analytical and critical approach. An attempt is also made to highlight the latest developments in comparative environmental law world.

This book has been divided into six parts. Part One introduces the subject followed by the facts and outcome of the Oleum leak. Part Three deals with the environment in general. The next Part is the heart of the present exercise where administration of environmental justice has been critically evaluated. Part Five deals with hazardous industries and legal control. The present work closes with the conclusion and recommendations to minimise the danger and maximise the safety requirement/against the industrial pollution.

The present work opens a new vision to the researchers in the field of law. It will be of great use to the students of environmental law, practitioners, judges, the managers of industries and all those who are involved in the industrial processes and interested in the protection of the environment.

I take opportunity to express my profound gratitude to Prof. C.M. Jariwala of Law School, Banaras Hindu University, Member, Commission on Environmental Law, Switzerland and International Council of Environmental Law, Germany, whose continuous motivation and guidance have contributed immensely in the completion of this work.

I have no words to acknowledge my debt of gratitude to my parents, brothers, specially my younger brother Prem and sisters for their consistent support and abiding faith. I am deeply obliged to my friends and in Particular to Mr. Vidya Sagar, for their active cooperation and support. Last, but not the least, I am extremely thankful to Mr. S.B. Nangia of Ashish Publishing House for the present publication and also for bringing out the book with a nice get up and that too within a very short period.

Vinod Shankar Mishra

Varanasi

Contents

Table of Cases

P

R

S

PART I : INTRODUCTION

Introduction

We love life and yet court death. That is paradox of the human predicament. The good that I would, I don't; the evil that I would not, that I do, 'says St. Paul.' The Mahābhārata observes: *Jānāmi dharmam na ca me pravṛttiḥ janamy adharamaṁ na ca me nivrttih,* 'I know the right, but I don't adopt it. I know the wrong, but I cannot abstain from it.' It is the crisis in the human soul which is reflected in the world. Unless we overcome it in the minds and hearts of men, there will be fear for the future.[1]

This 'paradox of human predicament' has caused the destruction of *earth's natural resources* and its life sustaining environment. The rapid and indiscriminate *industrialization* have become an effective tool for endangering human life, health and surrounding environment.

Industry is the backbone of development. It is considered as standard to weigh prosperity and civilization of a country. In the present time it is believed that the more industrious a nation is the more prosperous is the country. Such prosperous country calls itself a highly civilized nation but today though the developed nations are named as big powers but they have left far behind the nature and fruits of nature. If development continues with the present speed then today's nature will be left with *unnatural nature.* The *Rio Conference* (1992) has also drawn the attention of the world on the pollution caused by developed countries which call themselves 'the most industrial develop nation.' The West is now seriously thinking to put back the clock whether this will be possible only time will answer this query.

---

1 Excerpts from Chancellor's address at Special Convocation, 22nd January 1959 ; Ossasional Speeches and Writings, pp. 276-277 - cited in "Prajna" (Journal of the Banaras Hindu University) Vol. 34, Part - 2, 1989, 250-251.

There are many categories of industries but the prominent one which attracts the attention of the environmentalist is the *hazardous industry*. By hazardous industry we mean that process which may adversely expose the life of human beings, plants, animals and the environment in general. For example, chemical industry, atomic industry, fertilizer industry, etc. An industry having hazardous operations gives the maximum input to the industrialists. The industrialists of the hazardous operation have one point programme and that is make more profit with minimum cost. The selfish industrialists or the profit grabbing enterprises have put a question mark on the state of future's environment. Tens of millions of toxic and otherwise hazardous substance enter into environment every year as unwanted wastes. Managing and disposing of these hazardous waste is a great problem. An International Labour Organization study list has recorded no fewer than 6000 major hazardous installations in our country. We should not forget the fact that Industrial development around the world has contributed to 75 per cent of the global environmental problem. If one looks to the estimated *annual hazardous waste* generation in India as far as back 1984 it was about 0.3 million tons per year and it is expected that by 1992 it would be one million tons.

The Indian industrial sector has been ranked as the tenth biggest in world in gross industrial output. It has been proudly claimed that today India is in a position to compete with the outside world. The political *scenario* in the country has forced the government to initiate a number of measures to remove constraints on industrial growth and providing more congenial environment for the development. The liberalisation of industrial policy, it is claimed, has resulted in an average growth of 8 per cent. The *liberalised industrial policy* has given a lot of incentives to the local, national and multinational enterprises to work for more and more economic development of the country. India, 1991, published by Ministry of Information and Broadcasting, Government of India, while dealing with industry have concentrated mainly on economic development and surprisingly one finds hardly any reference to ecology and environment of India.

The impact of *indiscriminate industrialization* on human lives and ecology has been completely ignored by our policy makers. The biggest threat to human lives comes from the hazardous industries.

The *human settlements* are allowed to come up near such industrial zone. The relocation policies for toxic industries or warehouses have not been yet adopted. All the above factors have led us to 'worst industrial disasters' and they are equally shared by public and private sectors. They are exemplified by *Panipat disaster* related to *ammonia lead* (1992), *Grasim Mill in Gwalior* (1991), *oleum leak from Sriram Industries* (1985), and the most tragic one, *MIC leak from Union Carbide Plant in Bhopal (1984)*. However, neither the Government nor the industries have taken any lesson from these painful experiences.

Hundreds of toxic chemicals are produced, transported, handled in most cavalier fashion without any safety precautions. The government in the guise of liberalisation has been delicensing industries which have major safety implications. The most prestigious newspaper of the country *The Times of India*, calls it in its Editorial as 'prescription for disaster.'[2]

In fact, the last ten years (1980-1990) may be regarded as decade of '*hazardous pollution*' of our life sustaining environment. It was the *Bhopal gas tragedy (1984)* in which 3000 lives were lost, had aroused the "public consciousness" against hazardous industries and their acts of omission and commission. Although the new consciousness was mainly among the urban educated who attached the level of '*environmental conservation*' to an age old struggle.[3]

The early eighties also witnessed emergence of influential urban based non-government organization like *Environmental Service Group, The Centre for Science and Environment*, The *Kerala Shashtra Sahitya Parishad, Kalpavirksh* to name a few. In this connection, three document's one State of Environment published by Centre For Science and Environment have laid solid foundation for spreading Environmental literacy in India.[4]

All these factors have immensely contributed in raising the public consciousness against pollution created by hazardous

---

2. See Edit. of *The Times of India*, August 28th, 1992.
3. Singh, Shekhar; The Global Environmental Debate, *Indian Journal of Public Administration*, 1989 (July-Sept.), 363.
4. *The State of India's Environment*, 1982, *A citizen's report and state of India's environment* 1984-85. The *Second Citizen Report*. The State of India's Environment : The Third Citizen Report, 1991. (The Centre for Science and Environment.)

industries. Public spirited individuals have filed writ petitions in High Courts and Supreme Court for enforcement of fundamental right of citizens. Right to live in clean environment is one of such fundamental right which has been developed through bold and innovative interpretation of article 21.

M.C. Mehta, a senior advocate of Supreme Court has become synonymous with such public interest petitions. Our present study confines itself to one of such writ petition filed by Mehta.

The *Rio Conference* (1992) warning on the degraded state of affairs of present earth has thrown new challenges before to the academics to devote more time so as to avoid the existing industrial pollution in the world at large. It is time that we the law scholar should devote our attention to the pollution caused by hazardous industries. The present study is an humble attempt in this direction.

The present work concentrates on a hazardous industry because one knows about the ramification of such industries. The Chernobyl and the Bhopal mass disaster which are some of many experiences which will go in the history of world of environment for their devastating affect on human beings, animals, plants and rest of the other components of the environment. In order to understand the role of hazardous industries in the field of environmental pollution, a specific study is undertaken in the present work of *oleum leak from Sriram Industries*. This is an example to illustrate the ramification of *industrialization*. In the present study though the main focus is on *Mehta* case but wherever possible an attempt is also made to examine and highlight the related problems so that through *Mehta* the larger areas of industrialization and their impact on the environment could be exposed. In examining *Mehta's* problem the present work adopts an analytical and critical approach. An attempt is also made to highlight wherever possible in the comparative environmental law world.

This case is selected because the '*Sriram Industries*' was located in Delhi, India's capital, where the highest legislative, executive and judicial organs are continuously working. Sriram is a private and leading industry of India. It was a multi-product process which was situated in the heart of capital of India. A large population inhibits around the Sriram complex. A large number of workers are employed by Sriram. And the surprising thing is that they employed an environmental engineer to look after the problems of environmental pollution. The oleum leak from one of the establishment reached as far as Daryaganj, a most congested area in Delhi. There is a Central Board (Air and Water) to continuously supervise and monitor the problems of air and water pollution and yet the central area of the capital was badly polluted resulting in even death of a person.

In the present study an attempt is made to highlight the problems related to constitutional law, legal control of environmental pollution through specific environmental law and general related laws.

The present study is undertaken to highlight the merits and weakness of the existing environmental laws and the laws related thereto. An attempt is made to examine how best the executive worked in consonance with their fundamental duty to protect the environment. It is the function of the judiciary to balance the development and environment. Is the judiciary successful in this endeavour? Has the judiciary made any attempt to build up a friendly environment? There is not much renovation in judicial set up in India from the British *Raj* to *Swaraj*. The question which also has to be examined is whether the present judicial set up is in a position to balance the scale of development and environment or it requires some modifications or complete renovation altogether.

In order to study the above problems, an humble attempt is made in the present work under the six parts. Part one introduces the subject. Part two gives the bare outline of the facts of *Mehta*, the currents and cross-currents therein and the resume of the outcome of the *Mehta* case. Part three consists of three chapters. Each chapter deals with some basic problems in *Mehta* case. It includes the position of Sriram as the state, fundamental right and duty relating to environment and environmental education. Part four is the heart of present exercise. It consists of three chapters and deals with neo concept of absolute liability, administration of environmental justice and the remedial measure. Part five deals with hazardous industries and legal control. In the end, Part six, the concluding part enumerates *Mehta's* directions. The present work ends with conclusion and certain recommendations so that hazardous operation may not further pollute badly the Indian environment.*

---

* The present study is based on literature published *upto August 1992.*

PART II : FACTS AND OUTCOME

CHAPTER 1

*The Oleum Leak : The Facts**

The *Delhi Cloth Mill Ltd.* is a public limited company having its registered office at Delhi. It runs an enterprise called *'Sriram Food and Fertilizer Industries.'* This enterprise has several units engaged in manufacturing of caustic soda, chlorine, hypochloric acid, stable bleaching powder, super phosphate, vanaspati, soap, sulphuric acid, alum hydrous sodium sulphate, high test hypochlorite and active earth. All these units are situated in a single complex spread over approximately 76 acres. This area is surrounded by thickly populated colonies. The population within a radius of 3 km of this complex is approximately two lakhs.

This plant was commissioned in 1949 and it has strength of about 263 employees including executives, supervisors, other staff members and workers.

In March 1985, the attention of Parliament was drawn on the possibility of a major leakage of liquid chlorine from the caustic chlorine unit of Sriram. It was also pointed out that such accident would cause damage to the lives of thousands of workers and other persons living in the nearby areas. It may be pointed out that the Parliament also showed concern in this regard and the Minister for Chemical and Fertilizer stated on the floor of the Lok Sabha that Government of India was fully conscious of the problem of hazards from dangerous and toxic processes and assured the House that necessary steps for securing observance of safety standards would be taken early in the interest of workers and the general public. This apprehension is also corroborated by the various expert committees appointed by the government as well as by the court. The committees

* In this part only the preliminary facts are given. The detailed account of the oleum leak is narrated in other chapters under the relevant heads.

emphasized the danger of an accidental release of the chlorine gas which could affect the workers and community living in the vicinity of Sriram complex.

It was on 4th December 1985, a major *leakage of oleum gas* took place from one of the Units of Sriram. This leakage affected a large number of persons both amongst the workmen and the public in the vicinity including an advocate practicing in the Tis Hazari Court who, it was alleged, died on account of inhalation of oleum gas. This leakage occurred from the bursting of the oleum gas tank as a result of collapse of structure on which it was mounted and it created a scare amongst the people residing in that area. On 6th December, 1985, there was another minor leakage, and it took place as a result of escape of oleum gas from the joints of a pipe.

The immediate response of the Delhi Administration to these two leakages was that the District Magistrate passed an order under section 133(1) of the Criminal Procedure Code on 6th December 1985 directing Sriram, within two days from the issue of the order, to cease carrying on the occupation of manufacturing and processing hazardous and lethal chemical and gases including chlorine, oleum, super-chlorine, phosphate at their establishment in Delhi and further to remove such chemicals and gases within seven days from the said places and not to again keep or store them at the same place. The court directed Sriram to appear on 17th December, 1985, in the court of District Magistrate, Delhi to show cause why not an order of closing down Sriram's establishments should be enforced.

It may be mentioned that while these proceedings were going on before the court, the Inspector of Factories, Delhi in exercise of his power conferred under section 40 sub-section (2) of *Factories Act, 1948*, issued an order on 7th December 1985 against Sriram phohibiting them from using the caustic chlorine and sulphuric acid plants till adequate safety measures were adopted and imminent danger to human life was eliminated.

Soon thereafter on December 13, 1985. a show cause notice was issued by Assistant Commissioner (Factories) of the Municipal Corporation, Delhi calling upon Sriram to show cause as to why not an action for the revocation of its license be taken under section 430 (3) of the *Delhi Municipal Corporation Act, 1957*, for violating terms and conditions of license. Sriram by its letter dated 23rd December, 1985, showed cause against the proposed cancellation of its license but by an order dated 24th December, 1985, The Assistant Commissioner Factories directed the Sriram to stop the caustic chlorine plant with immediate effect. The result was that unless these two orders, one dated 7th December, 1985 and the other dated 24th December, 1985, were vacated or suspended, Sriram was not in a position to restart the plant.

Problems of Water Pollution

Since some of these plants of Sriram including the vanaspati plant were discharging effluent, Sriram was required to obtain consent for discharging of effluent from the Central Board under section 25 of the *Water Act, 1974*[1] and Sriram accordingly made an application for this purpose in the prescribed form. The Central Board passed an order on 19th April 1979, granting consent to Sriram to discharge effluent from their factory in the sewer, subject to terms and condition set out in the consent order. The consent granted to Sriram was renewed from time to time and the last renewed consent order was passed on 22nd July, 1985 an it was valid upto 31st December, 1985.

The Waste water effluent from these four plants used to be drained through one common terminal outlet. The complaint of the Central Board was that this combined effluent at terminal outlet never complied with the standard prescribed by the board.

The affidavit of Surendra Kumar, Senior Environmental Engineer of Sriram Company stated that the control technology was based on the recommendation of the Central Board. In view of this, Messers Dorroliver were selected by Sriram in consultation with central board for supplying of an effluent treatment plant. But, unfortunately, the plant of Messers Dorroliver failed to give guaranteed results. On 17th January 1985, the central board directed Sriram that Messers Krofta Engineering Company should be asked to set up a pilot plant based on dissolve air flotation technology in the Vanaspati plant. But despite the follow-up actions taken by Sriram the pilot plant was not commissioned by Messers Krofta Engineering Company. Sriram there upon in its

---

1. The Water (Prevention and Control of pollution) Act, 1974.

anxiety to comply soon with the standard prescribed by the central board, placed another order with Messers Patel Brothers of Bombay in June 1985 for supply of plant based on flotation technology. Messers Patel Brothers guaranteed to install and commission the plant by 31st December 1985. However, the affidavits, showed that there was further delay in the installation of this plant and this installation was now going to be completed by 28th February 1986, Meanwhile Sriram got installed at the terminal outlet a plant based on dissolved air flotation technology of Messers Krofta Engineering Company. The counter affidavit of Sir P.R. Ghare Khan, on behalf of central board, dated 13th January, 1986, showed that the representative of the central board verified that terminal plant had been installed.

The last renewed consent order of the central board dated 2nd July, 1985, was to expire on 31st December, 1985. Thereafter, Sriram could not operate the Vanaspati plant and discharge effluent unless and until the consent order was renewed. In view of this difficulty, the central board in its affidavit filed in this behalf by D.C.Sharma, Assistant Environmental Engineer stated that the board had no objection to grant temporary consent persuant to the provisions of the water Act on the condition that Sriram would comply with all the recommendation of various committees appointed by the court or otherwise and that such consent would be valid for a period of one month from the date of issue of consent order.

There was one more dimension of the problem of water pollution in this case and that was the most unsatisfactory state of affairs prevailing in the Delhi Municipal Corporation. The Nazafgarh area sewer was lying chocked since 1980 with the result that Sriram was not able to discharge its domestic effluent in the municipal corporation sewer and instead domestic effluent had to be discharged in the Nazafgarh drain thereby it adversely affected the water standard prescribed by the central board.

Problems of Air Pollution

The Central Government in consultation with the Central Board issued a notification under section 19(1) of the *Air Act*, 1981[2]

---

2. The Air (Prevention and Control of Pollution) Act, 1981.

notifying certain areas in Union Territory of Delhi as air pollution control area. The plants of Sriram were admittedly situated in the air pollution control area and the industries carried on by sriram also came within the schedule of industries specified in the Air Act. Sriram was, therefore, required to apply for a consent order from the central board under section 21 of the Air Act. An application was accordingly made by Sriram, on the basis of which a consent order was issued by the central board on 13th June, 1985, authorising Sriram to operate their plants in the Air pollution control area, subject to the conditions set out in the consent order. The consent order was concerning three plants of Sriram, namely sulphuric acid plant, superphosphate plant and power plant.

CHAPTER 2

*Currents and Cross-Currents**

(A) ARGUMENTS OF PETITIONER AGAINST PERMISSION TO RESTART THE PLANT

The petitioner who appeared in person submitted vehemently and passionately that the court should not permit the caustic chlorine plant to be restarted because there was element of hazard or risk to the community in its operation. He urged that chlorine was dangerous gas and even if utmost care was taken, the possibility of an accidental leakage could not be ruled out and it would, therefore, be imprudent to run the risk of allowing the chlorine plant to be restarted.

(B) ARGUMENTS SUPPORTING OPENING OF PLANT

The learned counsels appearing on behalf of *Lokhit Congress Union and Karmachari Ekta Union*, expressed themselves emphatically against the permanent closure of the plant. It was submitted that if caustic chlorine plant was not allowed to be restarted, it would not be possible to operate plant manufacturing the down stream products[3] and the result would be that about 4000 workmen would be thrown out of employment. They contended that since all the recommendations made by various committees had been complied with by the Management of Sriram and that the possibility of risk or hazard to the community had been considerably minimised or reduced to nil, Sriram may be allowed to restart their establishment.

---

* In this part only general currents and cross currents find place. For specific arguments and counter arguments see the relevant headings.

3. Down stream products such as sodium sulphate, hydrochloric acid, stable bleaching powder, superchlor, sodium hypochlorite.

(C) ARGUMENTS OF GOVERNMENT OF INDIA AND DELHI ADMINISTRATION

The learned Additional Solicitor General appearing on behalf of the Union of India and Delhi Administration stated that his clients were not withdrawing their objection to the reopening of chlorine plant but if the court was satisfied that there was no real risk or hazard to the community, it might make such order as it is deemed fit. However, strict condition should be imposed with a view to ensuring the safety of workmen and the people in the vicinity.

(D) ARGUMENTS OF SHRIRAM FOR REOPENING OF THE PLANT

The learned counsel of Sriram strongly pleaded that now all the recommendations made by various committees had been complied with by the management and every possible steps had been taken for the purpose of ensuring the complete safety in the operation of the caustic chlorine plant. The management of Sriram made it clear that they did not intent to restart immediately their plants manufacturing sulphuric acid, oleum, chlorosulphuric acid, sulpher phosphate and granulated fertilizer ferricalum and active earth. The only plants in respect of which Sriram sought the permission of the court to restart were the power plants and the plants manufacturing vanaspati and refined oil including its by-products and recovery which included plants like soap, glycerine and technical hard oil and caustic chlorine plants including plants manufacturing by-products such as sodium sulphate, hydrochloric acid, stable bleaching power, superchlor, sodium hypochlorite, and container works. These plants could not be restarted by the management of Sriram unless and until the caustic chlorine plant was allowed to be reopened, because hydrogen was needed for vanaspati and refined oil plant and hydrogen would not be available unless the caustic chlorine plant was put into operation. There was no real danger of escape of chlorine gas and even if there was leakage, it could be only of small quantity and such leakage could easily by contained and there was no reason for permanently closing down the above plant. It was further argued that the permanent closure of the caustic chlorine plant would result in not only in loss to the company but also in unemployment of about 4000 workmen and it would result in the non-availability of chlorine to Delhi water Supply undertakings and short supply of down stream products.

CHAPTER 3

The Verdict of the Court

The Supreme Court took note of the arguments of petitioners and Government of India and Delhi Administration but it opined that since all the recommendations of *Man Mohan Singh Committee, Nilay Chaudhary Committee* and *Dr. Sharma Committee* were complied with by the management in satisfactory manner, Sriram could be allowed to restart the caustic chlorine plant. The court came to the above conclusion in view of the following circumstances:

(*i*) The recommendations of Various Expert Committees including the last one appointed by this court, had been complied with by the management of Sriram, thus possibility of hazard or risk to the community was considerably minimised.

(*ii*) If plant was not allowed to reopen, it would throw about 4000 workmen out of employment and such a closure would lead to their utter impoverishment.

(*iii*) The Delhi water supply undertaking which got its supply of chlorine from Sriram would also have to find alternative source of supply and it was common ground that such source was available quite at a distance from Delhi.

(*iv*) The production of downstream products also would be seriously affected resulting in short supply of these products.

Though these considerations were taken into account still Supreme Court felt :

> It is none too easy task, for the decision either way may entail serious consequences. We have, therefore, reflected over the various aspect of this rather difficult and complex question

with great anxiety and care and taking an over all view of diverse considerations, we have with considerable hesitation, bordering almost on trepidations reached the conclusion that, pending consideration of the issue whether the caustic chlorine plant should be directed to be shifted and relocated at some other place, the caustic chlorine plant should be allowed to be restarted by the management of Sriram, subject to certain stringent conditions.[4]

1. *Conditional Order*

The conditions imposed by the court included :

(1) A committee of experts shall be appointed by the Supreme Court:

   (*a*) To monitor the operation and maintenance of the plants and equipments.

   (*b*) A fortnightly inspection of the plant would be carried out and a report to that effect would be forwarded to the court after the completion of inspection.

(2) There shall be personal or individual responsibility of operator and Head of caustic chlorine plant for safety devices.

(3) The Chief Inspector of Factories shall inspect the factory once in a week. The Inspector shall report to the court and Labour Commissioner about any default, deficiency remissness on the part of the management.

(4) There shall also be an inspection by representative of the Central Board for compliance of provision of water and air Acts.

(5) The management shall give an undertaking for the payment of compensation to the victims of any gas leak, and further, that it shall deposit security money in that regard.

(6) In order to further look after the safety arrangements, a committee shall be constituted consisting 3 representative of the 'Lokahit Congress Union' and 3 representative of the 'Karmachari Ekta Union'.

---

4. *M.C.Mehta vs Union of India* AIR 1987 SC 965, 075.

(7) The management shall display notice in Hindi and English on the gate stating the effects of chlorine gas on human body and its treatment.

(8) Sriram shall organise a training programme for every worker regarding the functioning of specific plant and also, conduction of refresher course at least once in six week with mock trials.

(9) Sriram shall install loudspeaker in factory for a timely warning in the event of leakage of chlorine gas and maintain proper vigilance.

(10) In view of the damage caused by the present oleum leak the management was required to deposit an amount of Rupees twenty lakhs and to furnish the Bank guarantee of Rupees fifteen lacs within a period of two weeks from today for compensation claims of victims of gas leak. The court categorically stated that in case the management did not abide by the above conditions, the permission granted by this court would stand withdrawn.

2. ***Conditions Modified***

Of the long list of conditions there were more conditions in respect of which modifications were sought by Sriram on the ground that their compliance would entail enormous operational and practical difficulties.

The first condition in respect of which modification was sought by Sriram, was relating to personal responsibility of operator and Head of the caustic chlorine plant for safety devices.

The Supreme Court opined that the Head of caustic chlorine plant and the officer who was placed in charge of each group of safety devices would be personally responsible but it excluded the poor operator from such responsibility of safety devices.

Second condition regarding which modification was sought by Sriram related to the personal responsibility of the Chairman, Managing Director and officers for the payment of the compensation to the victims of gas leak.

The Supreme Court allowed the original condition to continue. However, in case of officer who was acting as the Head of

Caustic chloride plant, the liability of such officer was limited to the extent of his annual salary with allowances. But such officer should be entitled to indemnify the loss by Sriram if he could show that the escape of gas took place as a result of act of God or vis major or sabotage or he had exercised all due diligence.

The last condition in respect of which modification was sought by Sriram related to the constitution of committee to look after the safety arrangement in the caustic chlorine plant.

The Supreme Court did not dilute this condition. However, it felt necessary to make certain clarifications:

The workmen of committee should give prior intimation to the officer-in-charge with at least half an hour notice before leaving the duty for inspection of safety measures of any department of caustic chlorine plant.

The Supreme Court reiterated that the permission granted by it to Sriram to open the caustic chlorine plant was subject to the conditions set out in the order dated 17th February 1986 as modified by this order. It also added that if for any reason, Sriram did not comply with any of these conditions and was, therefore, unable to reopen the caustic chlorine plant, it would be open to Sriram to restart the other plants in respect of which permission had been granted by the order dated 17th February 1986 so long as Sriram could do so without operating the caustic chlorine plant.

PART III : ENVIRONMENT IN GENERAL

CHAPTER 1

Is Sriram a 'State' Under the Constitution ?

(A) DEFINITION OF 'STATE' UNDER THE CONSTITUTION

Article 12 of Indian Constitution defines the term 'state' which includes the following :

1. The Government and Parliament of India *i.e.*, executive and legislature of the union.
2. The Government and the legislature of each state *i.e.*, executive and legislature of states.
3. All local or other authorities within the territory of India.
4. All local and other authorities under the control of the Government of India.

The controversy in Sriram is centered round on the interpretation of the expression 'other authorities'. This makes it necessary to know the developments in this area so as to better appreciate the present case.

The interpretation of the term 'other authorities' in article 12 has caused a good deal of uncertainty. Judicial opinions have undergone changes during the last two decades. Today the government performs a large number of functions because of the prevailing philosophy of social welfare state. The government, in order to perform certain functions, has to depend on natural persons or juridical persons. Some doubts have been raised as regards the character of autonomous bodies. This raises the question : Can autonomous bodies be regarded as 'authorities' under article 12 ? An autonomous body may be statutory body set up directly by a statute, or it may be non-statutory body, a body registered under the general law : such as under the *Companies Act*, the *Societies-Registration Act*, or *State co-operative Societies Act*, etc.

In *Rajasthan Electricity Board* vs *Mohan Lal*[1], the Supreme Court was called upon to consider whether the Rajasthan Electricity Board was an authority within the meaning of expression *'other authorities'*.

Bhargava, J, who delivered majority judgement, pointed out that the expression other authorities would include all constitutional and statutory authorities on whom the powers were conferred by law.

Shah, J, who delivered a separate judgement, agreeing with the majority, preferred to give a slightly different meaning to the expression 'other authorities'. He opined that authorities, constitutional or statutory, would fall within the expression 'other authorities' only if they were invested with sovereign power of state, namely the power to make rules and regulation which had the force of law.

In *Sukhdev* vs *Bhagat Ram*[2], Mathew, J, propounded a broader test. According to the learned Judge, the *concept of 'State'* had undergone drastic changes in recent years and today state could not be conceived simply as "Coercive machinery welding thunderbolt authority" rather it had to be viewed mainly as service corporation. The honourable Judge treated the public corporation as an instrumentality or agency of the "state". But this distinction between the public and private corporation is diluted by justice Alagiriswami, dissenting. According to the learned judge, there was no distinction between an ordinary company and a statutory corporation as both stood on the same footing, one derived authority from the Companies Act and the other from the Act which created the corporation.

Now coming to the test to determine whether a corporation established by statute or incorporated under the law, is an instrumentality or agency of the government.

---

1. AIR 1967 SC 1857.
2. AIR 1975 SC 1331.

Justice Bhagawati in *R.D. Shetty* vs *The International Airport Authority*[3], opined that it was not possible to formulate an inclusive or exhaustive test which would adequately answer this question. According to the learned judge,

> There is no cut and dried formula which would provide the correct division of corporation in to those which are instrumentalities or agencies of the Government and those which are not.[3a]

However, the court developed following tests :

> Institution engaged in matters of high public interest or public function were by nature of the function performed Government agencies. Thus existence of deep and pervasive state control may afford an indication that corporation was a state agency or instrumentality.
>
> Secondly, if the function carried out by the corporation was of public importance and closely related to governmental function, it would be relevant factor in classifying the corporation as an instrumentality or agency of the Government.
>
> Thirdly, if the department of Government was transferred to a corporation, it would be a strong factor supportive of this inference of corporation being an instrumentality or agency of the Government.

The question regarding the *status of non-statutory body* was finally clinched in *Ajay Hasia* vs *Khalid Mujib*[4], where a society registered under the *Societies Registration Act*, running a regional Engineering College sponsored, supervised and financially supported by the Government, was held to be an authority for the purposes of article 12. This decision brings in the purview of 'other authorities' even a registered society.

The rationale behind this decision was that when the Government acted through its instrumentality or agency of the corporation and juristic veil of corporate personality work for the

---

3. AIR 1979 SC 1628.
3a AIR 1979 SC 1628.
4. AIR 1981 SC 487.

purpose of convenience of the management and administration, it could not be allowed to obliterate the true nature of the entity. The Supreme Court took this stand because what was at stake was the fundamental right. The court's concern for safeguarding of the fundamental right is reflected when it pointed out,

> Constitutional guarantees should not be allowed to be emasculated in their application by narrow and constricted judicial interpretation. The court should be anxious to enlarge the scope and width of fundamental rights by bringing within their sweep every authority which is an instrumentality or agency of the Government or through the corporate personality of which the Government is acting, so as to subject the Government in all its myriad activities, whether through natural person or through corporate entities to the basic obligation of the fundamental right.[5]

In this case, the Supreme Court also set at rest the controversy as to the origin of a corporation. According to justice Bhagwati, speaking for the Court, what was necessary was,

> ....not as to how the juristic person is born but why it has been brought in to existence. The corporation may be statutory corporation created by statute or it may be a Government company formed under the companies Act, 1956, or it may be society registered under the Societies Registration Act, 1860 on any other similar stature.[6]

It would come within the ambit of article 12, if it was found to be an instrumentality or agency of state on proper assessment of the relevant factors.

The line of reasoning adopted by justice Mathew in *Sukhdev Singh Case*[7] and improved by justice Bhagwati had brought out a sea-change in the attitude of judiciary in interpretation of '*other authorities*' under article 12. The aftermath of the judicial dynamism is that now the Indian Statistical Society[8], Indian Council of

---

5. *Ajay Hasia* vs *Khalid Mujib* AIR 1981 SC 487, 492.
6. *Ajay Hasia* vs *Khalid Mujib* AIR 1981 SC 487, 494.
7. AIR 1975 SC 1331.
8. *B.S. Minhas* vs *Indian Statistical Institute*, AIR 1984

Agricultural Research[9], and many more public and private enterprises find a place in the expression 'other authorities'.

(B) POSITION OF SRIRAM AS 'STATE'

The industrial policy resolution, 1956 had classified industries into three categories. The first category belonged exclusively to the state. The second category comprised those state owned industries where private sector was allowed to supplement the 'state owned' industries by developing undertaking on its own or with state participation. And the third category included all the remaining industries. Schedule B to the resolution enumerated the industries. The policy resolution also stated eighteen basic industries of importance, whose planning and regulation was under the control of the Central Government. These basic industries included chemical and fertilizer industries.

The Industrial (Development and Regulation) Act, 1951, was enacted to carry out the objective of *Industrial Policy Resolution, 1948*. Section 2 of the Act declared that it was expedient in the public interest that the Union Government should take under its control the industries specified in first schedule. Chemical and fertilizers found a place in the first schedule under items 19 and 18 respectively.

The basic question in *Sriram* was whether Sriram was 'the state' under article 12 of the Constitution and, if so, then the provision of fundamental right would be applicable in the present case.

Arguments of the Applicant

It was argued on behalf of the petitioner that even private corporations manufacturing chemical and fertilizers could be said to be engaged in activities which were so fundamental to the society as to come within the governmental function. It was pointed out that Sriram was registered under the Industries (Development and Regulation) Act, 1951, its activities were subject to extensive and detail control and supervision of the Government. It was further contended that the Government had issued license to run the establishment and it had the power to prevent over concentration

---

9. *P.K. Ramchandra Ayer* vs *Union of India* AIR 1984 SC 541.

in particular region or over investment in particular industry by refusing license to a particular unit.

Section 18G of the *Industrial (Development and Regulation) Act, 1951* empowered the Government to control the supply, distribution, price, etc., of articles manufactured by the scheduled industry and section 18A authorised the government even to assume management and control of such industrial undertaking. It was found that affairs of the company were being managed in a manner detrimental to public. Section 18AA provided for exceptional provision to take over the industry even without any investigation.

Further, Sriram was required to obtain license under the Factories Act and was subject to the direction and order of the authorities under the *Water (Prevention and Control of Pollution) Act, 1974*, and the *Air (Prevention and Control of Pollution) Act, 1981*.

Counsel for applicants also pointed out that the sizable loans, lands and other facilities was granted by the government to Sriram in carrying on the industry, and as such, taking the aid of American '*State Action*' doctrine, it was argued that such private activity, if supported, controlled or regulated by state, might be so entwined with governmental activity as to be termed as 'State Action' and it would then be subject to same constitutional restraints as the state.

In view of the above pervasive control of the government, the petitioner took the stand that Sriram was covered by the expression "*other authorities*" under article 12 and thus it came within the expression 'the state' for the purpose of Part III of the Constitution of India.[10]

Arguments of Sriram Against the Application of Article 12

Counsel for Sriram pleaded not to bring within the ambit of article 12 private corporation. He contended that the control or regulation of a private corporation by the state was under general statutory laws such as Industrial (Development and Regulation) Act, 1951, and this was the exercise of '*police power*' by the state.

---

10. The following cases are cited to support its stand
    (*a*) *Sukhdev* vs *Bhagatram* AIR 1975 SC 1331.
    (*b*) *R.D. Shetty* vs *International Airport Authority*, AIR 1979 SC 1628.
    (*c*) *Ajay Hasia* vs *Khalid Mujib* AIR 1981 SC 487.

It was emphasized that '*state control*' must be of the type where management policies of corporation by necessity of prior approval of government or by sizable representation on the Board of Management or by any other mechanism was adopted. The counsel of Sriram also pointed out that American doctrine of 'State Action' was not suitable in Indian situation. He further stated that once an authority was deemed to be other authority under article 12, it was "State" for the purpose of all its activities and functions. Moreover, the American functional dichotomy by which some function of an authority could be termed as '*State Action*' and other private action, could not operate here.

The learned counsel also pointed out that those rights which were specially intended by constitution makers to be available against the private parties were so provided in the Constitution specifically such as articles 17, 23 and 24. Therefore, to so expand the article 12 so as to bring private corporation, in its purview, would be against the scheme of Part III relating to fundamental rights.

Opinion of Supreme Court

The Supreme Court traced the historical background of 'State Action'. The court pointed out that in America, since the fourteenth (14th) Amendment was available against state, the courts in order to thwart racial discrimination by private parties devised the theory of 'State Action' under which it was held that wherever private activity was aided, facilitated or supported by state in significant measure, such activity took the colour of 'State Action' and was subject to constitutional limitation of fourteenth Amendment. The principle behind the doctrine of state aid, control and regulation so impregnating a private activity as to give the colour of state action was of interest to us and that also to a limited extent to which it could be Indianized and harmoniously blended with our constitutional jurisprudence.

The Supreme Court rightly did not apply the American doctrine rather it opined,

> We in no way consider ourselves bound by American exposition of constitutional law and this is because according to the court, "social condition in our country is different."[11]

---

11. *M.C. Mehta* vs *Union of India* AIR 1987 SC 1086, 1097.

This is well demonstrated by the fact that in *Raman Shetty's* case,[12] this court preferred the minority opinion of Doughlas, J in *Jackson* vs *Metropolitan Edition Company.*[13] Again in *Air India* vs *Nargish Mirza*[14] this court preferred the minority view of *General Electric Company* vs *Martina Gilbert.*[15]

It is unfortunate that the Supreme Court missed an opportunity to decide the issue whether Sriram fell within the ambit of article 12. The court left the matter for the future occasion.

It is surprising or rather disappointing that the Supreme Court of India and its judges who are given the title of the champions of the fundamental rights have missed the high wave length in the present case. Even the liberty lover justices like Chief Justice Bhagwati in the present case failed to deliver justice by leaving the question open. Though in the present case the Supreme Court left the matter open but a microfine reading of the judgement would show that in the heart of heart the judges wanted to bring Sriram within the arena of *'other authorities'*. This is supported from some of the observation made by the judges when it was said that the purpose of expansion had not been to destroy the reason *d'etre* of creating corporations but to advance the human right jurisprudence or "Primarily to inject respect for human rights."[16]

The Supreme Court also rejected the apprehension of counsel of Sriram that inclusion of private corporation within purview of article 12 and subsequently subject the private corporation to article 21 would be a death blow to the policy of encouraging and permitting private entreprenual activity.

The Supreme Court took this apprehension as an approach of *Status quoist* who had no courage and strength to face existing realities and opined,

> wherever a new advance is made in the field of human rights, apprehension is always expressed by *status quoist* that it will

---

12. AIR 1979 SC 1628.
13. (1974) 42 Law ed 2nd 477.
14. AIR 1981 SC 1829.
15. (1976) 50 Law ed 2nd 343.
16. *M.C. Mehta* vs *Union of India* AIR 1987 SC 1086, 1097.

> create enormous difficulty in the way of smooth functioning of the system and affects its stability.[16a]

Similar apprehension was voiced when this court in Raman Shetty's[17] case brought a public sector corporation within scope and ambit of article 12 and discipline of fundamental right[18] and speaking against the apprehension the court rightly concluded :

> It is through creative interpretation and bold innovation that human right jurisprudence has been developed in our country to a remarkable extent and this forward march of human rights can not be allowed to be halted by unfounded apprehension expressed by *status quoist*.[19]

We may also bring support to this view that Sriram was 'the state' from various tests laid down by Supreme Court from time to time.

The *'relevant factors'* capable of changing Sriram into an instrumentality or agency of the state may be summarised as follow :

(*i*) The activity of producing chemical and fertilizer was of vital public interest.

(*ii*) Private corporation like Sriram manufacturing chemical and fertilizer could be said to be engaged in activities which were so fundamental to the society as to be labelled as governmental function.

(*iii*) Sriram was registered under the *Industries (Development and Regulation) Act, 1951*, its activities were subjected to extensive supervision of the government.

(*iv*) Under section 18A of the Industrial (Development and Regulation) Act, 1951, government had pervasive control to even assume the management and control of an industrial undertaking.

(*v*) Sriram was required to obtain license under the Factories Act and was subject to direction and order of authorities under the Act.

16a. *Id* at 1098

17. AIR 1979 SC 1628.

18. *M.C. Mehta* vs *Union of India* AIR 1987 SC 1086, 1098.

19. *Id.* at 1098.

(*vi*) Sriram was subject to extensive Environment Regulations under the *Water (Prevention and Control of Pollution) Act, 1974, Air (Prevention and Control of Pollution) Act, 1981*, and also the recent *Environment Protection Act, 1986.*

(*vii*) Sizable loans, lands and other financial assistance were granted by Government to Sriram in operating the industry.

(*viii*) Sriram was engaged in the activity which had the potentiality to invade the right to life of large section of the people.

(*ix*) Impact of pollution was the same whether it was by a private corporation or public corporation. So it was necessary to treat both on par for the purposes of fundamental rights.

The above points also go to support that the expression '*other authorities*' must be given wide expansion which would attract the activities of Sriram. This stand will allow Part III of the Constitution of India to withstand with strength attacks and counter-attacks on the fundamental rights.

CHAPTER 2

Fundamental Right and Duty Relating to Environment

(A) CONSTITUTIONAL PROVISION

The Indian Constitution is the first Constitution in the world which made provision for the protection of environment. It incorporated certain significant concepts in the fundamental law of the country through the *Constitution (Forty-second Amendment) Act, 1976.*[20] The obligation to improve the environment and to safeguard forest and wild life has been imposed on the state by article 48A, Part IV of the Indian Constitution. Article 48A says, "The state shall endeavour to protect and improve the environment and to safeguard the forest and wild life of the country." It uses the word 'environment' which means the aggregate of all the external conditions and influences affecting life and development of organs of human beings, animals and plants. Article 48A also separately provides for safeguard of forest and wild life.

These obligations are not to be enforceable through any court but society may act as people's court to pressurise the state to take effective action in this direction. Parliament may be law create watch-dogs over state obligation in article 48A. Another innovation of the 42nd Amendment Act is the insertion of article 51A(g) which specifically deals with the fundamental duty of the Indian citizen with respect to environment. It says, "it shall be the duty of every citizen of India to protect and improve the natural environment including forests, lakes, rivers and wild life and to have compassion for living creature."

---

20. See for a detailed discussion C.M. Jariwala, the Constitution 42nd Amendment and the Environment, *Legal Control of Environmental Pollution*. 1980, 7, ed. S.L. Agrawal.

There is no provision in article 51A(g) as to the enforcement of this duty but it cannot remain a mere pious obligation because firstly, the laws relating to environment provide for the enforcement machinery and secondly, the court while balancing the fundamental right chapter would take the help of the fundamental duties. However, the fundamental duty may be subject to the difficulty that the general masses in the country are ignorant of the environmental problems and it is too early to impose on every citizen of India a constitutional fundamental duty in this regard before making necessary arrangement for its mass education.

Every duty implies the existence of a correlative right. The incorporation of fundamental duty relating to environment does not find any specific fundamental right in the Constitution of India. However, the Indian judiciary has concretized a right in this connection through a bold and innovative interpretation of article 21. The *Maneka* wave length has opened doors for neo-fundamental rights. And in this dynamism, the right to live in a clean environment finds a place through the judicial constitutional legislation.

The first case where the Supreme Court recognised the right though not directly was *Rural Litigation and Entitlement Kendra* vs *State*.[21] In this case the question was whether extension could be granted to continue with mining operations. The court took the stand that mining activities caused ecological disturbance and violated the rights of people to live. According to the court, the hardship caused to lessees was a price that had to be paid for protecting and safeguarding the right of people to live in a healthy environment.[22] In this branch of law, the High Courts have been more dynamic.

In *Damodar Rao* vs *S.O. Municipal Corporation, Hyderabad*[23] the Andhra Pradesh High Court had to settle the question whether part of area earmarked under a development plan for recreational purposes could be acquired and used by two state agencies for construction of residential houses. The court held that the attempt

21. AIR 1985, SC 652.
22. AIR 1985, SC 652.
23. AIR 1987, AP 171.

to build houses in such an open space meant for recreational park was contrary to the law. In this connection the High Court opined that,

> .....It therefore, becomes the legitimate duty of courts as the enforcing organ of constitutional objectives to forbid all action of the state and citizen from upsetting the environmental balances.[24]

The court categorically decleared that the right to environment was a part of right guaranteed under article 21. It went on and point out that the slow poisoning by the polluted atmosphere caused by environmental pollution and spoilation be regarded as amounting to violation of article 21 of the Constitution.[25]

The judicial trend is further supported in *Attakoya Thanqal* vs *Union of India*.[26] The case relates to a plant for pumping up ground water for the purpose of supplying potable water in a group of Islands, in the Arabian sea.

The Kerala High Court[27] directed the Union of India, under whose administration the island territories come, to have a deeper scientific study into the question before the scheme was given a go by. Justice Shankaran Nair, held that the right to life in article 21 contains the right to have clean water. He observed, the right to sweet water and the right to free air, are attributes of the right to life, for these are the basic elements which sustain the life.[28]

(B) *OUTCOME* IN MEHTA

In *M.C. Mehta* vs *Union of India*,[29] the Supreme Court found an opportunity to examine the potentialities of the fundamental right to live so as to include the right to live in a clean environment. The M.C. Mehta[30] is a case specifically dealing with an activity

---

24. AIR 1987, AP 171, 178, 179.
25. AIR 1987, AP 171, 181.
26. 1990 KLT 580.
27. 1990 KLT 590.
28. *Ibid.*
29. AIR 1987 SC 965.
30. *Ibid.*

threatening the life of workers and of the public in general. The court pointed out that the case raised 'some seminal questions concerning the true scope and ambit of articles 21 and 32 of the Constitution.[31]

The court carved out the right to live in a healthy environment through reading article 21. The emergence of this right required the state to secure this right. In this wave length the court authorised the state to place restriction on carrying on of hazardous industrial activities. The Supreme Court, further, emphasized the need to locate hazardous industries to safer places and insisted on minimization of risk to the community by maximization of safety requirement.[32] The Supreme Court evolved certain conditions[33] so as to allow the hazardous industry. In the second order[34] the court modified some of the previous conditions. In the third[35] the court in order to further safeguard the above right, order, allowed the right to claim compensation for damages due to hazardous activities.

Thus, the *Mehta's* case further nourished the tender plant of the right to a clean environment and thus it gets a firm ground in the Indian soil. It has been forecasted that such nourishment to "the liberty plant will make India once again a beautiful world."[36] However, it may be mentioned that the right has to be balanced with development. If the said right is allowed its full vigour in all its totality then it would retard the development process and this in turn would be fatal to the interest of the country. In this regard Bhagwati, J, showed concern when he opined,

> When science and technology are increasingly employed in producing goods and services calculated to improve the quality of life, there is certain element of hazard or risk inherent in the very use of science and technology and it is not possible to totally eliminate such hazard or risk altogether. We cannot

31. AIR 1987 SC 965, p. 966.
32. *M.C. Mehta* vs *Union of India*, AIR 1987 SC 965, 981.
33. *Id* at pp. 978-980.
34. *M.C. Mehta* vs *Union of India*, AIR 1987 SC 982.
35. *M.C. Mehta* vs *Unior of India*, AIR 1987 SC 1086.
36. Arun Kumar Mishra, *The Indian Water Pollution Law : Restrospect and Prospect*, 154. An unpublished LL.M. Dissertation of the Banaras Hindu University, 1987.

> possibly adopt a policy of not having any chemical or other hazardous industries merely because they pose hazard or risk to the community. If such a policy were adopted, it would mean end of all progress and development. Such industries, even if hazardous, have to be set up since they are essential for economic development and well being of people.[37]

Thus, the right and development must move forward in cooperation with each other.

---

37. *M.C. Mehta* vs *Union of India*, AIR 1987 SC 965, 982.

CHAPTER 3

Environmental Education

(A) LEGISLATIVE PROVISION

In India, environmental law primarily derives its authority from three Acts, namely *Water (Prevention and Control of Pollution) Act, 1974, The Air (Prevention and Control of Pollution) Act, 1981 and the Environment Protection Act, 1986.* Though these enactments have not expressly used the word 'environmental education' but these legislations have sown the seeds for environmental education in India.

Section 16 or Water (Prevention and Control of Pollution) Act, 1974 in sub section (*d*), (*e*), (*f*) mentions ways and means to spread the environmental education which are as follows :—

(*d*) To Organise the training of persons engaged or to be engaged in programme of water pollution on such terms and conditions as the Central Board may specify.

(*e*) To Organise through mass media a comprehensive programme regarding the prevention and control of water pollution.

(*f*) To Collect, compile and publish technical and statistical data relating to water pollution and measures devised for effective prevention and control and prepare manuals, codes of guides relating to treatment of disposal and sewage and trade effluents and disseminate information connected therewith.

Similar provisions are incorporated in the Air (Prevention and Control of Pollution) Act, 1981 under Section 16 Sub-Section (*e*), (*f*), (*g*).

The environment protection Act, 1986 in Section 3 Sub-Section XIII, also makes provisions as to how to educate masses. These specific provisions put emphasis on the following aspects:—

XII. Collection and dissemination of information in respect of matters relating to environmental pollution.

XIII. Preparation of manuals, codes or guides relating to the prevention, control and abatement of environmental pollution.

It is interesting to note that three environmental Acts have almost similar provisions relating to environmental education. But it is pity that the Central Pollution Board and State Pollution Board or other law enforcing agencies have yet to take initiative to make these novel provisions operational. However, the Water and Air Acts have emphasised on training of persons engaged or to be engaged in programme of water or air pollution. The responsibility of carrying out this task has been placed on Central Board. It is painful to point out that the Central Board has not fulfilled this task and thus the environmental education seems to be a distant dream.

The establishment of a separate department of Environment in 1980 is an important contribution for the conservation of environment. This department oversees environmental problems, policies and programmes all over the country and has launched many projects and programmes in the areas of pollution monitoring the controlling environmental pollution, environmental impact assessment, natural living resources conservation and ecodevelopment. It has also been given the task of encouraging environmental researches, environmental education, training and public awareness and transmission of environmental information. An *environmental information system (ENVIS)* was set up in 1982 as a decentralised system. The ENVIS network with Department of Environment as its focal point presently consists of 10 ENVIS centres in diverse area of environment such as pollution control, toxic chemicals environmentally sound and appropriate technologies, energy and environment, degradation of waste.

(B) CONSTITUTIONAL PROVISION

Coming to the supreme law of the land, no specific right to know, right to information or right to education finds a place in our

Constitution. The *Universal Declaration of Human Rights, 1948* specifically provides for this right in article 19 along with freedom of expression.

According to *S.P. Sathe,*[38] right to know, has two dimension. In broader sense it means right to education, to learn etc. In the narrower sense it means right to have information. We are concerned with broader aspect that is right to know means right to education.

The *Prabhu Dutt*[39] and *Sheela Barse*[40] cases show Supreme Court was of the opinion that the right to information was not always co-extensive with right to freedom of speech and expression.

In *S.P. Gupta*[41] the Supreme Court was articulating right to open government which was based on the people's right to know. In that situation the court took the stand that the right to know existed independently of the right to freedom of speech. Further, right to know would also be a part of right to personal liberty protected by Article 21 of the Constitution.[42] The aftermath of *Maneka* Gandhi[43] was that the right to life and personal liberty taken together was expanded to include the right to education. In *A.V. Chandel* vs *Delhi University*[44], Justice Deshpande opened a new chapter in Article 21 of the Constitution of India. he opined that

> Once the state takes action by legislation or otherwise to make the right to education available to every eligible person the statutory right or statutory facility should stand in practice (if not in theory) on the footing of fundamental right.[45]

Thus the above discussion lead us to one direction that the right to education, the right to know and the right to information are gaining the constitutional recognition in India.

---

38. S.P. Sathe, Right to Know, 1991.
39. AIR 1982 SC 6.
40. AIR 1983 SC 379.
41. APR 1982 SC 149.
42. Right to Know: S.P. Sathe 1991.
43. AIR 1978 SC 597.
44. AIR 1978 Del 308. See also Mohini Jain vs State of Karnataka. AIR 1992 SC 1058.
45. *Id* at 311.

(C) JUDICIAL APPROACH AND OUTCOME IN *MEHTA*

1. The Initial Approach

As far back as 1986 in *Bombay Environmental Group* vs *Pune Cantonment Board*,[46] the right to information was successfully claimed by the petitioner to be a part of the fundamental right chapter. In this evolution the Rajasthan High Court also made an important contribution. In *L.K. Koolwal* vs *State*[47] the Rajasthan High Court while considering a public interest petition by a citizen seeking protection against the neglect of sanitation by the State, observed;

> A citizen has right to know about the activities of State... why the state is withholding information in such matter.[48]

Recently, the Supreme Court has recognised that the dissemination of information is the foundation of system in democratic polity and keeping the citizens informed is an obligation of the government.[49]

Supreme Court, accepted on principle the prayers made by petitioner, *M.C. Mehta*, for issuing appropriate direction to the district collector to ensure exhibition of two slides on environment and pollution in cinema halls and video parlours as a condition for issuing license for these establishment. The Supreme Court also directed the Doordarshan and All-India Radio to produce daily programmes with a duration of five to seven minutes with message on environment and a regular weekly programme on the subject. The court has further directed the education boards to take steps to enforce compulsory education on environment upto matriculation from the next academic year. It also directed the University Grants Commission to consider the feasibility of making environment a compulsory subject at every level in the college and university education.[50]

Thus the Supreme Court of India has been preparing a solid background for the environmental education in India. Once this

---

46. Appellate side writ petition No. 2733 of 1986. See *Environmental Activities* ed. Gayatri Singh and Madhusudan Rao, 1988, pp. 172-175.
47. AIR 1988 Raj 2.
48. *Id* at 4.
49. M.C. Mehta vs Union of India, AIR 1992 SC 382.
50. *Id* at 384.

home work is done than there can be an effective people's participation in the conservation of the environment.[51] In India today the sad story is that not much is done to allow participation of the people of India in the area of environmental protection.

2. **Outcome in Mehta**

In the first *M.C. Mehta* case, the Supreme Court of India forged ahead the area of environmental education when it required Sriram management. Firstly, to place in each department or section of the caustic chlorine plant as also at the gate of premises a detailed chart in English and Hindi stating the effects of chlorine gas on human body and informing the workmen and the people as to what treatment should be taken in case they are affected by leakage of chlorine gas. And secondly, to organise a training programme for every worker regarding the functioning of specific plant and also to conduct refresher course atleast once in six weeks with mock trials. And lastly, to further educate the worker, audio-visual programmes will be prepared by the management.

The above observations go to support the conclusion that the *M.C. Mehta* case made an important contribution in the field of environmental education. Now it is time that we implement the legislative and judicial directions so that we the people of India may become environmentally not only educated but also conscious to protect the environment for us and for the future generation as well. The environmental illiteracy in India is one of the causes, that has allowed the environmental legislative inaction and in turn has resulted in the Indian environment to further degradation.[52]

51. See P. Leelakrishnan, Public Participation in Environmental Decision-making in *Law and Environment*, Editors Leelakrishnan Chandrasekharan and Rajeev, 1992, 162.
52. See Prof. S.K. Agrawala, Environmental Pollution : Issues for consideration in *Legal Aspects of Environmental Pollution and its Management.* Ed. S.M. Ali, 1992, 14.

PART IV : ADMINISTRATION OF ENVIRONMENTAL JUSTICE

CHAPTER 1

Juridical Balancing : Problems and Solutions

(A) PROBLEMS

There should really be no antimony between law and justice, but sometimes, conflict do arise between procedural requirement and goals of justice. Justice is the end of any civil society and law is an instrument of justice : "Our founding fathers had a fighting creed, a democratic order of human pervasiveness, egalitarian richness and a living social justice. They envisioned a constitutional harmony in action for fulfilment of this revolution through the rule of low.[1]

Twentieth century has witnessed rapid progress in scientific and technological field. This revolution greatly influenced the social, economical and industrial structure of the society. These developments are responsible for the growth of a nation. Today the United States, Japan and other developed countries have made so much progress because of the scientific and industrial innovations. This is one side of the coin and the other part depicts a dark side including the environmental pollution. The more the development a country makes, the more adversely is the environment affected. How much development is permissible and what degree of environmental deterioration may be allowed are some of the complex issues of the industrialization in the twentieth century.

The developmental progress has brought in number of problems and the one with which we are concerned is the problem before the judiciary of balancing the developmental progress and environmental protection. The *Mehta* case, is one of the series of

---

1. V.R.K. Iyer : *Law* vs *Justice*, 1981, 9.

case law on environment where the Indian Supreme Court felt great difficulty in administering the environmental justice. And the critics of the Indian Supreme Court, without going in detail with the overall performance of the court, concluded that the court was 'not innovative and illequipped to deal with complex legal issues', not possess and pool skills, the fund of experience.[2]

The judiciary, while administering environmental justice faces the following problems :

1. Education of judges - lack of scientific and technological knowledge.
2. The conflict between development and environmental protection laws.
3. Role of expert in environmental justice.
4. Philosophy for Individual judges.

(B) THE SOLUTIONS

(*i*) Education of Judges

Indian courts have to adjudicate disputes involving pollution control, ecological disaster and to reconcile the balance between environmental protection laws and the development and have to decide the liability of giant or multinational corporation. This is not possible without new expertise and multi-disciplinary vision. If we look to the early education of judges, we will find that it does not give them required skill to decide the complex question involving highly industrial and technical matters. And it is because of this reason, Felix Frankfurter once observed that there are certain areas where the "judges are rarely endowed and which their education does not normally develop."[3] In view of this difficulty Justice Krishna Iyer has advocated for the judges, a grounding in environmental justice.

---

2. See the Affidavit of Mare S. Galanter, *Mass Disaster and Multinational Liability* : *The Bhopal Case*, Prep. by Baxi and Paul, 1986, 161. See also the contraversion of Dadachanji his Affidavit before Justice Keenan's U.S. District Court in *Inconvenient Forum and Convenient Catastrophe* : *The Bhopal Case*, Prep. by Baxi, 1986, 72; See also *Id* at 2 Baxi's Countercriticism.
3. V.R. Krishna Iyer : *Our Courts on Trial*, 1987, 9.

If we take the stock of the educational qualifications of the judges of the Supreme Court and the High Courts, we will find that in most of the cases the judges have the B.A. or B.Sc. degree along with the LL.B. degree. This shows that they have no education in the advanced scientific and technical studies. If may be pointed out that mere having degree is not enough but one has to be upto date with the developments in the present and also time to come. This is true in case of environmental science. What causes environmental pollution? What degree of pollution is permissible against development? What control mechanism is reasonable? What consequences of pollution will the judiciary take cognizance of? What remedy the court will make available? These are some of the questions which the judiciary is confronted with while handling the environmental litigation.

In *M.C. Mehta* the Supreme Court felt the necessity of scientific and technical knowledge to decide the environmental cases. And in order to help the court in this difficult task, the Supreme Court advocated the necessity of neutral scientific expert as an essential input to inform judicial decision making.[4]

Education for judges is essential not only to update their knowledge of law and to create spirit of inquiry and innovation but also to inject a new empathy with the community and wider view of science is necessary for a competent judge. Justice Krishna Iyer further points out that judicial business management skill must be developed if our backward methods and archaic practices, which currently inflict enormous inconvenience, waste of time and money, delay and discontent and unscientific techniques, are to quit us.[5]

Thus it may be said that it is prerequisite for judges to have scientific and technical background before deciding the dispute relating to environmental pollution.

(*ii*) An Unbiased Mind

The last but not the least problem affecting the administration of environmental justice is the philosophy of individual judges. That is, they are guided by their own philosophy in delivering justice to the affected parties or group. Is judicial objectivity

---

4. *M.C: Mehta* vs *Union of India*, AIR 1987 SC 965, 981.
5. V.R. Krishna Iyer, *Our Courts on Trial*, 1987, 94.

possibly and can we have unbiased judges are some of the questions raised against the judicial approach. It is said :

> Judges are not hermits. They do have their set of ideas, attitudes and passions. A judge who is 'magestic and serene seeking inspiration as it from the Stars... with the detachment of sear, the voice soft and glorious, meant to fill the exalted cathedral... citadel of justice itself... is only a large assumption of fine fancy.[6]

The personality of judges is shaped by many factors like cultural, social, family background, etc. which play an important role in shaping the personality of a judge. Justice Cardozo rightly said :

> The decisions of the judges are considerations of social philosophy. He is influenced by inherited instincts, traditional believers, acquired convictions and conceptions of social needs. He must balance his analogies, history, custom, sense of right and all the rest and adding a little here and taking out a little there must determine as wisely as he can which weight shall tip the scale.[7]

Justice Holmes used to say a judge should always be conscious of the 'felt necessities'. And what are those felt necessities of our time? The felt necessities of our time are protection and preservation of environment from indiscriminate exploitation of natural resources and pace of rapid industrialization.

Justice K.H. Singh has castigated[8] the government for its failure to implement the environmental protection laws properly but he has also exhorted the commoner to follow law:

No law or authority can succeed in removing pollution unless the people cooperate.[9] Justice E.S. Venkatramaiah has also adopted the same approach and is in agreement with Justice K.N. Singh that "life, health and ecology have an edge over the other issues."[10]

---

6. V.R.K. Iyer, *Law* vs *Justice*, 1981, 72.
7. *Id* at 67.
8. *M.C. Mehta* vs *Union of India*, AIR 1988 SC 1115, 1121.
9. M.C. Mehta vs Union of India, AIR 1988 SC 1037, 1047.
10. *Ibid.*

Similarly Justice A.P. Sen and S. Natrajan are also champion of environment protection and conservation.[11] These judges have tilted the balance in favour of environment. If one looks to the separate judgement of Justice K.N. Singh in the tanneries case[12] he will find that the honourable Judge preferred the protection of river Ganga against a large number of tanneries of Kanpur. The river Ganga is considered by the Hindu as their mother Goddess and they cannot think of Ganga being polluted. Justice Singh, it seems, says Prof. Jariwala, was moving in this philosophical and emotional wavelength. On the other hand judges like Justice Misra, Dutta and Oza have mostly favoured the developmental side.[13]

The very first order in the present case shows the learned 'Judges' adopted a rigid attitude to save the environment even at the cost of development. If the first order was the final order then it would have caused unemployment, non-availability of certain chemicals and the essential commodities. It was only gradually that the temper of the judiciary cooled down and Sriram was allowed to operate its establishment under certain conditions. Justice Bhagwati's concern to save the environment is also reflected in his subsequent judgement which says:

> Preservation of environment and keeping the ecological balance unaffected is a task which not only government but also every citizen must take.[14]

In *M.C. Mehta* vs *Union of India*[15] Justice Bhagwati has evolved a new principle of tortious liability known as 'absolute liability' to meet the felt necessity of the twentieth century. In the instant case the learned judge has also shelved the procedural formalities and treated a letter' as a 'writ petition' to prevent the danger to the health of community and the environment in general. His philosophy got unanimous support from Justice Rangnath Misra, Justice G.L. Oza, Justice M.M. Dutt and Justice K.N.Singh.

---

11. *U.P. Pollution Control Board* vs *Modi Distilleries*, AIR 1988 SC 1128.
12. *M.C. Mehta* vs *Union of India* AIR 1988 SC 1037, 1047.
13. The present analysis of the philosophy of individual judges is taken from C.M. Jariwala. The Direction of the Environmental Justice in India; A Critical Appraisal of the 1987 case law; 1991, 34 (unpublished paper).
14. *Rural Litigation Kendra, Dehradoon* vs *State of U.P.* AIR 1987 SC 359, 363
15. AIR 1987 SC 1086

The above discussion shows that the justice is not a straight jacket, a mathematical calculation or a balancing scale. Somewhere it is said that the judiciary administers justice blind folded eyes. But the above data shows the direction towards justice influenced by the philosophy of individual judges. The pro or anti-environmental approach would keep the scale of environmental justice changing with the change in the Bench. In this uncertain situation what is necessary is that while administering the environmental justice the judges must keep in their mind that the justice is for the maximum good of the maximum number otherwise a rigid pro or anti-environmental approach would be suicidal to the social interests.

(*iii*) A Balanced Approach

Another problem in the administration of environmental justice is related to the conflict between development and the protection of environment. It is true that modernization requires technologically advanced industries which are essential for all round development of a nation. Development is the barometer to measure the progress in civilization of a country. The more the development the more progressive will be a society. Today there are few developed nations who want to be more and more developed on the other hand, there are developing countries who want to follow the footprints of the developed nations. In this race the underdeveloped states also do not lag behind. Thus there is a mad rush for development all around the earth.

The basic concept 'development' is highlighted by UNESCO.[16]

> Development must to designed, even at the humble level, as a process of ensuring the advancement of man through his own endeavours.

Development for what, for whom and by whom? Paul Harrison poses the problem scientifically; he observed:

Development for what purpose? Development towards what end? Every one is agreed on one point; that development must mean the eradication of absolute poverty and hunger and want. But after that, what? The field is wide open. It may be that the benefits

---

16. Paul Harrison, *The Third World Tomorrow*, 41.

of this type of rapid growth led by modern sector will eventually trickle down. But eventually can mean many generations.[17]

Industries are a modern tool for development. The parochial, regional and political consideration are deciding factor in location of industrial projects in India rather than environmental factors.

Justice Krishna Iyer has suggested that legal technology of development must ensure considerable participation of the people at various levels in the decision making and policy formulations operation and execution of programmes and projects tuned to Indian condition. The known how of development sometimes called 'appropriate technology' must be indigenous although we may adapt even adopt what is congenial to our condition.[18]

Emphasizing the role of 'law' as companion of the development process the learned judge has pointed out that law is aninalienable companion and must figure as a prominent member of developmental sisterhood of sciences and social sciences.[19] The 'law' should be responsive to the people's need to meet the changing requirement of 20th century.

Although laws relating to environment have not been much effective in providing a balance between the developmental process with the protection and preservation of environment.

Now coming to the Indian judiciary in many cases, the Supreme Court maintained a fine balance between developmental process and protection and preservation of environment. However, there were cases where courts favoured either developmental process or environment as the goal of justice.

Society for Protection of Silent Valley vs *Union of India*[20] is a case where the environmentalist, journalists and experts raised their objection against Construction of dam on Kuntipuzha river in Kerala's Palghat district.

---

17. *Id* at 16, 342.
18. Justice V.R. Krishna Iyer. *Environment Pollution and the Law*, 1984, 52.
19. *Id* at 8.
20. O.P. No. 2949 and 3025 of 1979. The Kerala High Court decided the case on *2nd January 1980*. The case had not been reported in any Law reports in India. See M.K. Prasad *"Silent Valley Case: An Ecological Assessment"* in Law and Environment, ed. P. Leelakrishnan, 1992, 117.

The immediate gain was the industrial development of the present area. Unfortunately the Kerala High Court dismissed the writ petition seeking protection of the environment. The court pointed out:

> It is enough to state that we are satisfied that the relevant matters have receive attention, before the government decided to launch the project. There has been no non-advertence of the mind to the silent aspect of the project. We are not to substitute our opinion and notion on there matters for those of government.[21]

The same approach was evident from granting permission to start an oil refinery or railway yard not far away from historical monument.[22]

In *Krishna Gopal* vs *State of M.P.*[23] a factory manufacturing glucose sline was permitted to be located in a residential complex, which caused public nuisance with noisy bolders used round the clock. Here the judiciary leaned in favour of development in preference to the environment.

The developmental wave length gets support in M.C. Mehta[24] when Chief Justice Bhagwati pointed out:

> We cannot possibly adopt a policy of not having any chemical or other hazardous industries merely because they pose hazard of risk to the community. If such a policy were adopted, it would mean end of all progress and development.[25]

Justice Bhagwati had emphasized the need of such industries for our economic development. But it is submitted that the above approach has been discarded even in the West where the governments are concerned with how to put check on the developmental process. The Hare Ram Hare Krishna wave in the United States have shown an apathy towards the developmental goals. Though we cannot order for the complete closure of all the

---

21. *Id* at 125.
22. Darryl D, Monte, *Temples or Tomb* 1986, Chapters 6, 7
23. 1986 *CriLJ* 396.
24. AIR 1987 SC 965.
25. *Id* at 981. See also *Sachidanand Pandey* vs *State of W.B.* AIR 1987 SC 1109, where developmental process was favoured.

industrial establishments yet we have to remember that we must leave, if not clean, a liveable environment for the present and generation to come. The challenge as to how to strike a balance between the competing claims of conservation and development is formidable indeed and has yet to be met effectively by planners, administrators, engineers and law makers of the country.[26]

We may sum up by saying that Indian judiciary has leaned in favour of development and has ignored the 'possible catastrophe' which might result from environmental degradation. Of course, individual justices have tried to tilt the balance in favour of environment protection and preservation.

(*iv*) An Expertised Environmental Justice

The most complicated problem is of the mass data, complexity of law fact issue, absence of formal legal research to analyse those data and absence of any statistics on similar points relating to environmental pollution and in particular an industrial disaster. All the above deficiencies have made difficult for the Indian judiciary to deliver fair, just and reasonable environmental justice.

Justice Krishna Iyer echoing the same view, has pointed out that judiciary does not have the equipment for social research or laboratory studies when confronted with large scale lawlessness in field of environmental pollution and industrial production. He has further explained that violation by noxious discharges, non-compliance with welfare measures and mafia defiance of human rights of humbler, weaker sector of society, demand careful handling if the remedy is not to aggravate the malady.[27]

It has been often seen in Indian High Courts and Supreme Court that judges tend to rely on expert opinion of either plaintiff or defendant. Of course, this reliance on expert submission of either parties is not out of favour on procedural niceties but due to lack of its own investigative machinery of commission which could advise them about technical and scientific aspect of the environmental cases.

---

26. Prof. (Kr) Gulam A. Khan, Environmental Pollution:Debate and Challenges, *Legal aspect of environmental pollution and its management*, 1992, 4. S. Musharraf Ali.
27. V.R. Krishna Iyer, *Our Courts on Trial*, 1987, 64.

The opinion of scientific and technical expert in administration of environmental justice has become an essential part of decision making process. In *Vincent* vs *Union of India*[28] and *Shivarao* vs *Union of India*[29], the Supreme Court declined to decide the issues before it for want of technical knowledge.

According to justice Krishna Iyer task technique gap has led the Indian courts to flounder on environmental issues. His further observation deserves our attention which says:

> The techniques, tools and technician continue as in British days but the task to be executed has been transformed. This task technique gap or hiatus between means and ends is at the root of justice crisis, as of other crisis.[30]

Gujarat High Court[31] had taken a lead to fill up this gap. The High Court constituted a commission headed by Prof. Upendra Baxi to investigate into the report on the working of the Surat Textile Workers so as to take considerate judicial action after survey operation.

Prof. Baxi had expressed his warm appreciation to High Court for appointing him as the Chairman of the action research group to assist the depressed, disadvantaged and depressed strata of Indian society. The opening words in his report are meaningful and deserves our attention :

> Only the cooperative endeavour of thinking humanity and suffering humanity which thinks will lead us to the attainment of constitutional socialism.

The power to appoint commission is an inherent power of the Supreme Court under Article 32 and High Court under Article 226 of the Constitution. Since the commission is constituted by eminent person in the desired field, the report of commission is *prima facie* evidence of facts and data collected. The report of the commission is supplied to the parties to rebut the contents of report through an affidavit. The Court decides the case taking into account the affidavits filed by the parties.

---

28. AIR 1987 SC 990.
29. AIR 1987 SC 952.
30. V.R. Krishna Iyer: Our Courts on Trial, 1987, 93.
31. *Id* at 62 and 63.

On a petition[31a] for direction to the municipal authority of city of Jaipur to solve the city's acute sanitation problem, the Rajasthan High Court appointed a commission to report on the insanitary condition in various parts of the Jaipur city. The commission discharged this function to the utmost satisfaction of the court.

There may be occasion when the commission may exceed the limit of the assigned task and may submit such subjective report which may be remotely connected with its assigned task. The Bombay High Court had to face such a situation in *Sergiocarvallo* vs *Sate of Goa.*[31b] The petition challanged the development permission for a beach resort on the Goa's coast. Since work at beach site had commenced, the court appointed a commission to investigate and report on the extent of construction. The commission exceeded its assigned task and commented on legality of structures. The high court expressed its dissatisfaction and rejected the subjective portion of the report.

Role of Expert in M.C. Mehta

In the *M.C. Mehta* case there was leakage of the oleum gas. The oleum gas was one of the ingredient in manufacturing varieties of goods in Sriram's Industries. This affected the workers, passers-by and those who lived in the vicinity of Sriram complex. It is alleged that one person died of the above leakage. Sriram pleaded that the goods produced in the industrial complex were essential to the people of Delhi and the public in general. Any adverse action on the industrial process would cause not only unemployment but also spending of foreign exchange to import similar goods from outside in case of scarcity. Thus, in the present case the court has confronted with the problem of balancing environmental protection *vis a vis* development. In order to help the court to arrive at a judicious balance, it requested the government and also on its own appointed committee after committee.

Since *M.C. Metha* case is a path finder in environmental technology and justice, the constitution of committee, their role in assisting the court and their recommendation need mention so as to assess the role of environmental expert in the administration of

31a. *L.K. Koolwal* vs *State of Rajasthan* AIR 1988 Raj-2.

31b. Writ Petition No. 367 of 1988.

environmental justice. A summary of the constitution and recommendations of various committees in the *M.C. Metha* case is given below :

There were six committees, constituted by the court, Government of India and the petitioner. Two committees were appointed before leakage of oleum gas, by the Government of India. One by the Labour Ministry and the other by the Delhi Administration to make a report about the potential problems likely to arise in future from the Sriram plants.

Two committees, which were appointed before oleum gas leak were :

Firstly, the Dr. Slater Committee ; and secondly, the Man Mohan Singh Committee.

1. **The Dr. Slater Committee**

This committee was constituted by the Ministry of Labour, Government of India. Dr. Slater was an expert of 'Technica' a firm of consultant scientist engineers of United Kingdom. The committee sent preliminary report. Since this committee was appointed before oleum gas leak, and its report was preliminary in nature. Thus the Supreme Court did not rely on this report.

Recommendation of the Dr. Slater Committee

(*a*) Dr. Slater had in last part of his report pointed out that inspection of chlorine plant revealed 'a worrying state of affairs'.

(*b*) The plant was liable to be 'classed as major hazard facility by applying most of the currently accepted definition.

(*c*) It did not measure up to the responsibilities incumbent upon operation of such plant to safeguard both public and employees as far as reasonably practicable.

(*d*) The authorities should consider constraining its activities to protect the public and employees.

(*e*) He concluded by observing that 'relocation is the only practicable long term option which would guarantee the complete removal of the community risk.

2. **The Man Mohan Singh Committee**

This committee was the result of deliberation in Indian Parliament (some time in the month of March 1985) on possible leakage of chlorine gas from caustic chlorine plant of Sriram. The Delhi Administration constituted an Expert Committee consisting of Man Mohan Singh, Chief Manager, IPCL Baroda, as Chairman and three other members to go into the existence of safety and pollution control measures covering all aspects such as storage, manufacture and handling of chlorine in Sriram.

The report of this committee was a detailed one, dealing exclusively with caustic chlorine plant's safety and measure thereto. The Supreme Court gave weight to this report.

Recommendations of the Man Mohan Singh Committee

(*a*) Man Mohan Singh Committee also observed towards the end of its report that 'total elimination of risk to the community *i.e.*, human population from toxic plant/hazardous industries located in close proximity was improbable.

(*b*) The probability of risk could be immensely reduced if the plant was run with adequate precaution.

(*c*) Man Mohan Singh Committee after considering the safety requirement observed that function of Sriram in the present location was not desirable.

(*d*) Man Mohan Singh Committee pointed out various drawback in the structure and design of caustic chlorine plant and also its maintenance and operation and made detailed recommendations which in its opinion needed to be strictly and scrupulously carried out to minimise the risk.

(*e*) Man Mohan Singh Committee opined that it was only where concentration of chlorine in the air was between 40 to 60 parts per million (PPM) that exposure for 30 minutes would be dangerous to the life.

Four committees were appointed after oleum gas leak. These were :

(*i*) The Seturaman Committee ;

(*ii*) The Agarwal Committee ;

(*iii*) The Dr. Nilay Choudhary Committee ; and

(*iv*) The Dr. Sharma Committee.

(*i*) Seturaman Committee

This Committee was constituted by Lt. Governor of Delhi under the Chairmanship of N.K. Seturaman and four other experts member to go into causes of spillage of oleum and the after effects of the leakage of oleum gas.

This committee's report primarily dealt with safety procedure in the sulphuric acid plant from which oleum gas leaked.

Recommendations of the Seturaman Committee

(*a*) Seturaman Committee pointed out in para 108.1 of its report that Sriram factory 'is certainly a perennial source of hazard to the community.'

(*b*) These hazard could not be completely eliminated but could be minimised by strict compliance of safety regulations.

(*c*) Seturaman Committee pointed out that the machinery and equipment in some of these plants were old and worn out.

(*ii*) Agarwal Committee

Persuant to liberty given by the Supreme Court, the petitioner (M.C. Mehta) appointed an Expert Committee, known as the 'Agarwal Committee' consisting of Dr. G.D. Agarwal, Professor T. Shivaji Rao and Sri Purkayastha. This committee pointed out various inadequacies in Sriram plants and opined that it was not possible to eliminate hazard to the public so long as the plant remained at the present location.

Recommendations of Agarwal Committee

(*a*) Agarwal Committee opined that chlorine manufacturing unit could not be even reasonably safe when located in proximity of densely populated area.

(*b*) It suggested that the only practicable solution was to relocate the chlorine plant at least 10 km away from the urban limit of densely populated areas with adequate safety measures.

(*c*) Agarwal Committee in its report stated that concentration of chlorine in the air above 25 parts per million was recognised by Occupational Safety and Health Act (USA).

(*iii*) Dr. Nilay Choudhary Committee

The Supreme Court appointed an independent team of Expert Committee under the Chairmanship of Dr. Nilay Choudhary, Dr. Aghoramurty, R.K. Garg and others as member, asking the committee to submit is report on following three points :

(*a*) The question of giving permission to restart the plant in the present state and condition ;

(*b*) The question of the safety measures to be adopted before the operation of the plant ;

(*c*) The question of non-installation of safety devices in plant by Sriram.

This committee visited the chlorine plant on 28th December 1985 and after considering reports of other committees, submitted 14 recommendations to the court for its compliance by the management of Sriram.

(*iv*) Dr. Sharma Committee

The Supreme Court made an order on 31st January 1986. The assignment was carried out by Sri Man Mohan Singh, Sri Garekhan and Dr. Sharma.

The report was submitted on 3rd February 1986.

The report showed that barring construction of a shed on space where filled cylinders were to be kept, all the recommendations made in the reports of Man Mohan Singh Committee had been complied with by the management of Sriram.

The committee pointed out :

(*a*) The hydraulic test carried out by 'Messers Nike Associates', Bombay, a firm recognised by Chief Inspector of Factories, Bombay as competent person to take up the responsibility of testing, examining and issuing certificate in respect of pressure vessels, also established that all the five tanks had an adequate capacity of withstanding pressures.

(*b*) The authorities wanted a hydraulic test to be carried out once again by Regional Testing Centre, Okhla, Management got a certificate issued by R.H.T.C. that all the five tanks were strong enough to withstand pressure of 375 dsig for 30 minutes duration.

(*c*) The insistence of committees that not more than 140 filled chlorine cylinder should be stored and report showed that this limitation had been accepted by the management of Sriram.

(*d*) The committee also insisted that a 'mock-drill' with a view to ensuring whether there was specifically trained group to handle any chlorine leakage in emergency and the committee stated in its report that mock-drill was found to be satisfactory.

All the expert committees were unanimous in their view that by adopting proper and adequate safety measures the element of risk to the workmen and the public could only be minimized but it could not be totally eliminated.

To sum up there was unanimity with respect to the operations of Sriram industrial complex that it was an hazardous establishment and it was highly objectionable to have such a complex in a thickly populated area. The committees felt once such industrial complex was permitted in such a locality then the only thing that could be suggested was 'maximum precautions' in its operation. Thus the direction of the committees report was maximum precaution and minimum degradation of the environment.

Looking to the role of the various committees in the *M.C. Metha* case, it may be said that 'experts' are inevitable to aid and advice to the court in the administration of environmental justice. This raises another problem whose help as an expert the court should take whether the experts hired by the plaintiff, defendant, the government or the court's own experts. In case of experts of the plaintiff or defendant, the experts will sing in the tune of the person who hire them and it for this they are paid. The government's experts will try to plead the developmental side. And the expert for the court would be in a better position to help the court in arriving at a judicious result. In view of the above problems, the Supreme Court in the present case rightly advocated for a neutral scientific expert as 'an essential input to inform judicial decison making.'[32] The Supreme Court deliberately used the word 'neutral', to emphasize the impartiality of such expert so that the vested interests do not prevail on the balancing scales and simply to protect their self-interest.

Emphasizing the need for scientific and technical expert, the Supreme Court pointed out that in this case several expert committees were appointed to inform the court as to what measures were required to be adopted by management of Sriram to safeguard against the hazards or possibility of leaks or explosion, causing pollution of air and water etc., and how many safety devices against this hazard or possibility existed in the plant were necessary. But in spite of so many high powered committees manned by well renowned experts in the field, the court, instead of being educated, "was confused" on the technological aspects ; in the present matter. It is surprising that in spite of the recommendations of high powered and top experts committees, the court still felt helpless. Who is to be blamed is a question which needs a detailed research. Is it because the judges are not ready to such exposures ? In view of the above difficulties, the Supreme Court suggested that the better course would be to take help of the neutral experts. But this solution is not so easy. How to determine who is a neutral adviser ? Where to find him ? In this connection the court also realised that there was no independent and competent machinery to generate, gather and make available the necessary scientific and technical information at the present. Moreover, such an exercise, according to the court was difficult and unsatisfactory.

Keeping these circumstances in mind, the Supreme Court came down with the recommendation that the Government of India may set up an ecological science research group consisting of independent professionally competent expert in different branches of sciences and technology who would act as an information bank for the court and also to the government department. The Supreme Court also pointed out that cases involving issues of environmental

32. *M.C. Mehta* vs *Union of India* AIR 1987 SC 965, 981.

pollution, ecological destruction and conflicts over natural resources were coming up for adjudication and these cases involved assessment and evaluation of scientific and technical data,[33] and therefore, there was an urgent need of the involvement of experts in the administration of justice.

(*v*) Establishment of an Environmental Court

Public consciousness[34] has contributed immensely in furtherance of administration of environmental justice in India since late nineties. Environment protection and preservation has acquired a new social dimension.

Indian courts are heeding the plea of environmentalist and conservationists to protect and preserve the environment through judicial intervention. In Bombay courts, alone over five thousand ecology oriented cases are reported to have been pending for a long time. There are also cases pending in the Supreme Court and other high courts. Thus the litigation relating to environmental pollution is increasing by leaps and bound. It has necessitated the requirement of separate court to deal exclusively with cases relating to environment and ecology.

At times the courts felt it difficult to handle the environmental litigation. In the *M.C. Mehta* case, the court had to take the help of the experts to apprise them of the problems. It has also been a subject of criticism that the courts have failed to balance the scale of environmental justice and development. It has also been pointed out that the present juridical process does not fit in with the environmental dispute settlement machinery. Thus, in India a dialogue started as far back as 1980 to have a separate machinery to deal with the environmental litigation. When Justice Desai became the Chairman of the Law Commission of India, he invited the law academics to discuss and debate on the necessity of tribunal to settle disputes of special nature. The need for an environmental court was first felt in the celebrated case of *M.C. Mehta* vs *Union of India*.[35] In this case, the Supreme Court drew the attention of the

33. *M.C. Mehta* vs *Union of India* AIR 1987 SC 965, 981.
34. Public spirited individual like environmental (Supreme Court) Lawyer, M.C. Mehta who has filed a number of petitions to protect the health and ecology of people.
35. AIR 1987 SC 965, 982.

Government of India towards the growing number of cases relating to environmental pollution and ecological destruction. Therefore, the Supreme Court suggested to the Government of India that it is desirable to set up an environmental court on the regional basis with one professional judge and two experts drawn from the 'Ecological Science research group'[36] keeping in view the nature of the case and expertise required for its adjudication. It also mentioned that there would be, of course, a right of appeal from the decision of the environmental court.

Although the concept of environmental court is noval one, but the Supreme Court has not cared to spell out the power and function of such court and judges. The courts' suggestion raises multifacet problems. What will be the status of these professional (scientific and technical) expert ? Can they be considered as judges ? From where these experts will be drawn ? In which branch of study the expert will be drawn ? From which stream will the judge come ? Should the judge member possess a special qualification ? How much time the court must take to settle the matter ? What judicial remedies will be available ? Can the presence of the experts will balance the scale ? This raises another problem, suppose the opinions of two experts contradict with each other then the professional judge would be in a very embarassing position as to which opinion he must support. He would be accused of biased against the professional whose advice has not been taken into consideration. It is desirable that such court may be constituted by one professional expert and two professional judges.

The Environmental Court Bill, 1990

The then Minister of Environment, Mrs. Maneka Gandhi, vehemently advocated for an environmental court. This line of thinking was carried forward by the present Minister of Environment and Forests who has been greatly concerned with the state of affairs of the Indian environment. The concern for the environmental justice at the top level activised Parliament to give statutory recognition to the long felt need of the environmental court. The drafting of the environment court Bill was started in December 1989. The Bill was intended to be similar to the

36. *Id* at 982.

consumer protection Act, 1986, under which the consumer courts have been set up. The constitution of environmental courts is the brain child of former Chief Justice of India, Justice P.N. Bhagwati who made a strong plea in *M.C. Mehta*[37] for separate court to deal with environmental issues and he was entrusted with the task of preparing the draft bill for the environmental court.

The bill envisages the constitution of a national and regional environmental court to deal with the environmental litigations. The Bill provides that the national environmental court and the state environmental courts are to be headed by a president and other members of whom atleast 1/3 are to be judicial members and the remaining to be drawn from the environmental experts pool, a body specially provided for, by the Bill.

Environmental Commission

An Environmental Commission has been envisaged under the Bill which shall select the president and judicial member of the national environmental court and environmental courts of states in consultation with the respective Chief Justices.

This Commission is also given power to constitute an environment pool, comprising minimum of 24 members, all of whom to be expert on environmental matter. The Commission's other duties include :

(1) To establish a legal cell to advise environmental and public interest group ;

(2) To provide free legal aid to the victims ;

(3) To carry out enquiries and investigation in the matters of environmental degradation ;

(4) To formulate and develop environmental law.

The formality of accompanying affidavit with the letter, petition of application addressed to the commission, have been discarded.

The environmental court could also employ the services of environmental commission, central or state pollution boards, environmental and social scientist etc., to enquire and prepare a report on the cases pending before the court.

---

37. AIR 1987 SC 965.

Environmental court has to give opportunity of hearing against the report to any party and then pass an order to dispose off the case.

The claim for compensation for the alleged injury done by environmental pollution should be made within three years from the date of such injury.

The controversial provision of environmental court bill, 1990 was related to the notification to environmental court and also publication of notice in two national and one regional newspapers regarding any project and its impact on the environment. It had been alleged that the bill had given unnecessary power of interference to environmental court and environmental group.

A provision was made for the constitution of the environment fund to carry out work outlined in the Bill and was to be administered by Central Government.

Relief under the Bill

The concept of 'public interest litigation' had been incorporated under the provision of the 'Bill'. The environmental court will either issue direction or orders granting injunctional relief to victims after receiving petition, application or letter complaining environmental degradation.

Non-compliance with any direction or order passed by the National Environmental Court or Environmental Court of any State of Union Territory was deemed to be an offence, punishable with imprisonment which was to be extended upto a period of 9 years or with a fine upto Rs. 1000 or both.

An Appraisal of the Bill

1. The environmental court would fulfil the 'felt necessity' of neutral scientific and technical expert to assist in decision making process. The environmental court is better suited to the regional or the grass-root level. There are cases where the grass root judiciary felt enormous difficulty in administering the environmental justice.[38] This problem to a certain extent could be minimised through a new set up suggested above.

38. *Ratlam Municipality* vs *Verdhichand and other*, AIR 1980 SC 162.

2. Environmental court will save time and money of the litigants and will provide immediate relief in the form of damages as the proposed bill requires a case to be disposed off within specified time limit of six month, giving no scope for indefinite prolongation of the legal battle.

3. The constitution of environmental court is criticised on the ground that it is costly exercise and it will retard the economic development and would cause bottle-neck to the process of development.

The criticisms is not based on solid facts. The case law on environment shows that judiciary has favoured economic development *vis-a-vis* environment.

It is unfortunate that Ministry of Environment has put the Environment Court Bill, 1990 in cold storage. Noble provision, which could have become harbinger of environmental peace and development has yet to see the light of the day. In fact, even in the absence of environmental courts, the High Courts and the Supreme Court are open to any conscentious citizen under the concept of public interest litigation. The *M.C. Metha* case[39] is an example of this kind.

In the present century when there is a demand for a world[40] in general, and European Environmental Court[41] in particular the time is ripe to have Indian environment courts.

(*vi*) New Procedural Innovations

The Fifth Amendment of U.S. Constitution lays down *inter alia* that "no person shall be deprived of his life, or liberty or property, without "due process of law". This clause has been the most significant source of judicial review in the U.S.A. The word 'due' is interpreted as measuring just, 'proper' or 'reasonable', according to the judicial view. Due process has two aspects :

---

39. *M.C. Metha* vs *Union of India* AIR 1987, SC 965, 981.
40. *The Times of India* : Feb. 19, 1992 See A Staff Reporter item on "World Environment Court likely Soon". See also Environmental Court : Need of the Hour—Vijay Chandra Tennet *Supreme Court Journal*, 1991, p. 36.
41. Philip Sands, European Community Environmental Law : Legislation, the European Court of Justice and Common Interest Groups. 53 *Mod. L. Rev.* 1990, 685.

substantive due process envisages that substantives provisions of law should be reasonable and not arbitrary. Procedural due process envisages a reasonable procedure, *i.e.*, the person affected should have fair right of hearing which includes four elements ; notice, opportunity to be heard, impartial tribunal and an orderly procedure. Under the concept of due process, the courts become arbiter of both substantive and procedural provisions in law.[42]

In India article 21[43] has acquired a new dimension after the Maneka Gandhi case[44]. Justice Krishna Iyer has observed :

> True, our Constitution has no 'due process' clause or V Amendment (of American Constitution) but in this branch of law, after Cooper and Maneka Gandhi's case the consequence is the same.[45]

Under article 32 of Indian Constitution, the Supreme Court is free to devise any procedure for the enforcement of fundamental right. The Constitution makers deliberately did not lay down any particular form of proceeding for the enforcement of fundamental right nor did they stipulate that proceeding should confirm to any rigid pattern or straight jacket formula because they knew that in a country like India where there is so much of poverty, ignorance, illiteracy, deprivation on rigid formula for enforcement of fundamental right would become self defeating.[46]

In *S.P. Gupta* vs *Union of India*[47] the Supreme Court recognised the revolution taking place in judicial process and opined that,

> the court has to innovate new methods and device new strategies for the purpose of providing access to justice to large masses of people who are denied their basic human rights and to whom freedom and liberty have no meaning.[48]

---

42. M.P. Jain : *Indian Constitutional Law*, 1987, 577.
43. Art. 21 is similar to U.S. V Amendment. it reads "No person should be deprived of his life or personal liberty, except according to procedure established by law".
44. AIR 1978 SC 597.
45. Sunil Batra vs Delhi Administration, AIR 1978 SC 1678.
46. *Bandhua Mukti Morcha* vs *Union of India*, AIR 1984 SC 807.
47. AIR 1982 SC 149, 105.
48. *Id* at 185.

This quest of Supreme Court to develop a new tool for social justice led to the invention of 'public interest litigation' or social action litigation.[49] It allows, a public conscious individual or individuals having an independent locus to file a suit under the extra-ordinary writ jurisdiction of the Supreme Court and the High Courts.

In *S.P. Gupta* vs *Union of India*[50] the Supreme Court gave a clear picture of the public interest litigations and the individuals right to sue. The court observed that under the traditional law the basis of entitlement to judicial redress was personal injury to property, body, mind or reputation arising from violation, actual or threatened of the legal rights of the person seeking such redress pointing out the individual's inability and emphasizing at the same time for the protection of his interest. The court held that where a legal injury was caused to a person or to a determinate class of persons and such person was by reason of poverty, helplessness of disability or socially or economically disadvantaged position unable to approach the court for relief, any member of public could maintain an application.

The Supreme Court also pointed out that it must not be forgotten that,

> procedure is hand maiden of justice and cause of justice can never be allowed to thwarted by any procedural technicalities. The court would, therefore, unhesitatingly and without slightest qualms of conscience cast aside the technical rules of procedure in exercise of its dispensing power and treat the letter of public spirited individual as a writ petition and act upon it.[51]

Similar opinions were expressed in *People's Union for Democratic Rights* vs *Union of India*[52] and *Bandhua Mukti Morcha* vs *Union of India*[53].

But the Supreme Court hastend to add that individual who moves the court for judicial redress in cases of this kind must be

---

49. See also Prof. S.K. Agrawal, Public Interest Litigation—*A Critique*, 1985.
50. AIR 1982 SC 149.
51. *S.P. Gupta* vs *Union of India*, AIR 1982 SC 149, 185.
52. AIR 1982 SC 1473.
53. AIR 1984 SC 802.

acting bonafide with a view to vindicating the cause of justice and if he is acting for personal gain or private profit or out of political motivation or other oblique consideration, the court should not allow itself to be activised at the instance of such person and must reject the application at threshold whether in the form of a letter or in form of regular petition filed in the court.

Justice Bhagwati in *Bandhua Mukti Morcha* vs *Union of India*[54] opined that a letter addressed to Supreme Court by person acting *pro bono publico* can be treated as writ petition. Such letter can be legitimately regarded as 'appropriate proceedings.'

Justice R.S. Pathak and Justice A.N. Sen opined that communication or petition invoking the jurisdiction of the court must be addressed to the entire court, that is Chief Justice and his companion judges.

Rejecting this contention, the Supreme Court in third *M.C. Mehta case*[55], opined through Chief Justice Bhagwati;

> We must not forget that letter would ordinarily addressed by poor and disadvantaged persons or by social action groups who may not know the proper form of address. They may know only a particular judge who comes from their state and they may address letter to him.[56]

Thus the Court reinforced its earlier stand that even if a letter was addressed to an individual judge of the court, it should be entertained, provided, of course, it was by or on behalf of a person in custody or on behalf of a woman or a child or a class of deprived or disadvantaged person.

The Supreme Court has made elaborate procedure to entertain such letters as petition. A public interest cell has been established in the Supreme Court. This cell is entrusted with all letters addressed to the court or to the individual justices. The staff members attached to this cell, examines the letters and it is only after scrutiny that the letters are placed before the Chief Justice and under the direction are listed before the court.

---

54. AIR 1984 SC 802.
55. AIR 1987 SC 1086.
56. *M.C. Mehta* vs *Union of India* AIR 1987 SC 1086, 1089.

The Supreme Court also felt that insistence on having affidavit with such public interest petition would frustrate the entire object and purpose of epistolary jurisdiction. Because poor and disadvantaged persons and social action groups would not be able to approach the Supreme Court.

The Supreme Court had also expressed its anguish over growing criticism of public interest litigation and pointed out that there was some misconception in the minds of some lawyers, journalist and men in public life that public interest litigation was unnecessary cluttering up the tiles of the court and adding to already staggering arrears pending in courts but that, could not be any reason for denying access to justice to the poor and weaker section of the society. Time had now come when the courts must become the courts of the poor and struggling masses of this country.[57]

The Supreme Court has aptly remarked that;

> ... It is true that there are large arrears pending in the courts but that can not be reason for denying access to justice to the poor and weaker section of the community. No state has right to tell its citizen that because a large number of cases of the rich and well to do are pending in the court, we will not help the poor to come to the courts for seeking justice till staggering load of cases of the people who can afford is disposed of.[58]

In *Municipal Council Ratlam* vs *Virdichand* and *others*[59] expressing his concern for protection and preservation of environment justice Krishna Iyer had observed:

> If the centre of gravity of justice is to shift, as the preamble to the Constitution mandates, from the traditional individualism of *locus standi* to the community orientation of public interest litigation, these issues must be considered. In that sense these issues before us between the Ratlam Municipality and the citizens of a ward, is a path finder in the field of people's involvement in the justicing process.[60]

---

57. *People's Union for Democratic Rights* vs *Union of India* AIR 1982 SC 1473, 1476.
58. *Id* at 1478.
59. AIR 1980 SC 1622.
60. *Id* at 1623.

The court further stressed the need to widen the scope of the rule of *locus standi* to embrace all interests of public minded citizens or organizations.[61] Thus, there was relaxation in rigidity of traditional procedural law in environmental cases also.

Thus 'public interest litigation has become a modern tool to provide environmental justice to suffering masses and it acts as protector of 'natural environment'.

In this connection, *Rural Litigation and Entitlement Kendra, Dehradun* vs *Union of India*[62] deserves our attention. In this case indiscriminate mining operations in lime stone quarries had disturbed the ecological balance of the Dehradun Mussoorie Hill area. The Supreme Court, recognising this case as first of its kind relating to environment, ordered for the closure of mining operation. The court showed its concern not only about the state of affairs in the Mussoorie Hill area but also its long term effects on the people and also on the Indian environment. The Supreme Court also observed.

> These (natural) resources are permanent assets of mankind, are not intended to be exhausted in one generation. Preservation of environment and keeping the ecological balance unaffected is a task which not only governments but also every citizen must take.[63]

In the first *M.C. Mehta*[64] case, a public interest litigation was filed under article 32 seeking for closure of Sriram Food and Fertilizer Industries engaged in manufacture of hazardous process. The present public interest litigation enabled the petitioner to move the highest Court of India not only to ask for closure of Sriram Establishment but also for damages. Whatever may be the results but the Supreme Court had the opportunity to pronounce judgment on such issues.

There was one more dimension of the present case and that was M.C. Mehta, the senior advocate of the Supreme Court who initiated the present proceedings and also proceedings in

---

61. *Fertilizer Corporation, Kamgar Union* vs *Union of India*, AIR 1981 SC 344.
62. AIR 1985 SC 652.
63. AIR 1987 SC 359, 363.
64. AIR 1987 SC 965.

subsequent environmental case law. This was appreciated by the Supreme Court and was evident from the concluding para of the judgement where the court observed:

> The petitioner had rendered signal service to the community by bringing this public interest litigation and producing considerable material on the issue arising in the litigation.[65]

The Supreme Court also pointed out that without this public interest litigation there would had been no improvement in the design, structure and quality of the machinery in the caustic chlorine plant nor would any proper and adequate safety devices and instrument had been installed nor would there had been any pressure on the management to observe safety standards and procedure.

The Supreme Court as token of its appreciation of the work done by the petitioner, directed the Sriram to pay Rs. 10,000 to the petitioner by way of cost. This gesture of Supreme Court reflected its willingness to compensate even those who initiate the proceedings.

In the third *M.C. Mehta* case,[66] the court held that right guaranteed under Art. 21 of the Constitution meant right to live in a pollution free environment. And to protect this right, the court allowed an application under public interest litigation for compensation against harm done by leakage of oleum gas.

It is gratifying to note that in the present case the Supreme Court provided legal aid to the victims of gas leak. It has conferred locus on the Delhi Legal Aid and Advisory Board to file an action on behalf of sufferers in the appropriate court for claiming compensation.[66a]

'Speedy trial' was another tool to provide justice to the victims of gas leak. In this case,[67] the Supreme Court directed the Delhi High Court to nominate one or more judges as might be necessary

---

65. *M.C. Mehta* vs *Union of India* AIR 1987 SC 965, 982.
66. AIR 1987 SC 1086.
66a. See also V.R. Krishna Iyer, Environmental Protection and Legal Defence, 1992, 54.
67. *Id* at 66.

for the purpose of trying such action so that they might be expeditiously disposed of.

It may be mentioned that the public interest litigation wave has not left the high court untouched. They have invoked writ jurisdiction under article 226 and significantly contributed in protection of environment through public interest litigation.

In *L.K. Koolwal* vs State of Rajasthan,[68] the Rajasthan High Court allowed a petition of the citizens of Jaipur for the preservation of sanitation in the city.

Kinkari Devi vs *State of Himachal Pradesh*[69] is another public interest litigation where the Himachal Pradesh High Court directed the closure of mining activities dangerous to the environment.

Thus, the public interest litigation has played a significant role in protecting and preserving environment. It is also instrumental in devising innovative method to provide instant relief to the victims of environmental pollution.

To sum up in the words of Justice Krishna Iyer :

> The democratic adaptation of judicial technology to meet the new challenges, with their ideological implications, has come to stay and can not be wished away. It has Constitutional roots, popular roots, popular backing and Supreme court support, although its birth Pangs and infant ailment still afflict it.[70]

(*vii*) Environment Tribunal

In the *M.C. Mehta*[71] the Supreme Court suggested an alternative forum for establishing an expert body to administer environmental justice. This has attracted the attention of Parliament and on 5th August 1992, the Government of India introduced the Environment Tribunal Bill[72] in Parliament.

The Bill was introduced by present Minister of State for Environment and Forest Mr. Kamalnath to implement the decisions

---

68. AIR 1988 Raj 2.
69. AIR 1988 HP 4.
70. V.R. Krishna Iyer *'Our Courts on Trial'*, 1987, 55.
71. AIR 1987 sc 965.
72. Bill No. 133 of 1992.

taken at the United Nations Conference on Environment and Development held at Riodde Janerio in June 1992.

Salient Features of The Environment Tribunal Bill, 1992

The Bill provides that the Central Government shall by notification, establish a tribunal to be known as the National Environment Tribunal. The principal Bench of the Tribunal will be located at New Delhi and such other places as the Central Government may specify. The Bill empowers the Central Government to distribute business among the Benches and specify the matter which may be dealt with by each Bench.

The Tribunal shall consist of a Chairman and such number of Vice-Chairmen, Judicial members and technical members as the Central Government may deem fit. However, each Bench shall consist of one judicial member and one technical member.

The primary qualification for a person to be appointed as Chairman, Vice-Chairman and judicial member of the tribunal is that, "the said person is or has been or is qualified to be a judge of a high court alongwith administrative experience of certain years."[73] A technical member must possess adequate knowledge, experience, capacity to deal with administrative, scientific or technical aspects of problem relating to environmental pollution.

The Chairman, Vice-Chairman judicial and technical members shall be appointed by the President in consultation with Chief Justice of India. In order to appoint the judicial and technical member, the Bill provides for a Selection Committee to be appointed by the Central Government. It shall consist of the Chairman of the Tribunal acting as ex-officio Chairman of the Selection Committee and ex-officio members of the committee would be secretary to the Government of India in the Ministry of Environment and Forest, Secretary to the Government of India in the Ministry of Law and Justice and Company Affairs, Director General, Council of Scientific and Industrial Research. The Bill also makes a place in the committee for an Environmentalist to be nominated by the Central Government.[74]

---

73. Section 11(1), (2), (3), (4), (5)
74. Section 11(7).

The judicial and technical member shall be appointed on the recommendation of this selection committee.

The Bill excludes the jurisdiction of any court or other authority and invests the tribunal of any jurisdiction, power or authority to entertain any application or action for claiming compensation. The Bill also empowers the tribunal with the same jurisdiction, powers and authority as the collector has in respect of matter specified in the Public Liability Insurance Act, 1966.[75]

An application for claim of compensation may be made by any person who has sustained injury, the owner of the property to which damage has been caused or by any legal representative of the deceased in case of death or by any authorised agent of such person or owner of such property.[76]

The novel provision of the Bill is the recognition of right of representative body or organization functioning in the field of environment to move the tribunal on behalf of the depressed and deprived class of citizens.[77] However, such body must be recognised by the Central Government, State Government or any other local authority.[78] This requirement would not allow the fake persons to move the tribunal to protect their private interests.

The Bill also makes provision that the complainant shall deposit a fee not exceeding one thousand rupees but it exempts the poor person whose annual income is less than one thousand rupees, the representative body or organization mentioned above.[79]

The bill imposes a limitation that no application for compensation shall be entertained unless it is made within 5 years of occurrance of accident.[80]

The Bill seeks to impose liability to pay compensation in certain cases on the principle of 'no fault'. It imposes liability on 'owner' of an enterprise where death or injury or damage to any person (other than the workmen), environment and property has

75. Section 5(3).
76. Section 5(1) (a to d).
77. Section 5(1)(c).
78. Section 5(1)(e).
79. Section 5(4).
80. Section 5(5).

resulted from an industrial disaster. In such a case the claimant need not prove any wrongful act, neglect or default of any person. The tribunal is empowered to apportion the liability for compensation on a equitable basis amongst those responsible for such activities.[81]

The decision of the tribunal is to be taken by the majority of the members.[82]

An award made by the tribunal under this Act shall be executable by the tribunal as a decree of Civil Court and for this purpose, it shall have all powers of Civil court.[83] The Tribunal could recover the money as arrear of land revenue which an owner has failed to deposit as directed by the tribunal.[84]

The Bill also makes provision for appeal against the award of the tribunal to the Supreme Court on one or more grounds specified in section 100 of Civil Procedure Code, 1908.[85] Every appeal shall be preferred within a period of 90 days from the date of award. The said limitation of 90 days may be relaxed in case if the appellant succeeds to show that he was prevented by sufficient cause from preferring the appeal within the said period.

There is no provision for appeal against an award or other award made by the tribunal with the consent of the parties.[86] The Bill also imposes a penalty for failure to comply with the order of tribunal which may extend to three years imprisonment or with fine of Rs. 10 lakhs or with both.[87] This liability will be attracted in case any person who was directly in-charge and was responsible too, the company for conduct of its business as well as the company itself. No action shall lie if the act was committed without knowledge and with due diligence.[88]

The Director, manager secretary or other officer of the company shall also be deemed to be guilty of an offence if the

---

81. Section 3(3).
82. Section 22
83. Section 24(1).
84. Section 24(3).
85. Section 24(1).
86. Section 22(2).
87. Section 26.
88. Section 27(1).

offence was committed with their connivance of negligence.[89] However, the Central Government may exempt an owner from the operation of this Act, namely the person who has sustained injury, State Governments, corporation owned or controlled by Central of State Government or any other local authority.[90]

The Bill seeks to vest the Central Government with unlimited power to determine the nature and categories of the officers required to assist the tribunal in discharge of its function.

An Appraisal of the Bill

The Environment Tribunal Bill, 1992, is a step forward in providing an alternative forum for delivering 'just, fair and reasonable justice.

The Bill has imposed an absolute liability' on the owner of an enterprise in case of an industrial disaster. In such cases, the claimant need not prove any wrongful, neglect or default of any person. Thus the Bill has in fact, accepted the diluted form of 'absolute liability' propounded in the *M.C. Mehta* case. It is interesting to note that the public liability Insurance Act, 1991 and the present legislation follow the 'no fault' principle in cases of payment of compensation. It means that an individual has two forums for claim compensation. These two forums exclusively deal with cases relating to environmental pollution.

The Bill provides the apportionment of compensation among those who are responsible for the pollution on equitable basis. This provision ensures a fair deal to every section of society.

The Bill has incorporated the basic philosophy of Public Interest litigation by granting an access to representative body or organization to file an application for compensation on behalf of victim. Perhaps, this is first legislation granting a statutory recognition to the Public Interest litigation. The Bill exempts the poor person and the non-governmental organisation from depositing 'mandatory feel' for filing such an application.

In the *M.C. Mehta* case, several expert Committees were appointed to assist the court on scientific and technological aspect

89. Section 27(2).
90. Section 4

of environmental pollution. In this case the Supreme Court strongly advocated for a neutral scientific expert as "an essential input to inform judicial decision making. This suggestion of the Supreme Court is now transformed in the prospective legislative exercise. The present Bill makes a provision for a 'technical member', in every Bench.

The Bill has ensured speedy justice by incorporating a provision for an appeal to the Supreme Court only. This will avoid further recourse from one court to another and would reduce the plethora of pending cases.

The Bill seeks to vest the Central Government with unlimited power to determine the nature and categories of the officers required to assist the tribunal in discharge of its function. It is the basic principle of administrative law that justice should not by only done but manifestly and undoubtedly be seem to done. There is a likelihood of bias in assistance from the officers to determined in cases of claims for compensation against the public servants or public enterprises. It is suggested that the tribunal should be given independent power to determine the nature and categories of the officers required to assist it.

The Central Government has been vested with unlimited discretionary power in granting an exemption from the liability in such cases it is necessary that certain guidelines should be prescribed to control the arbitrary exercise of the power.

The limitation of 5 years for filing an application to claim compensation may be applicable in the ordinary cases of injury but pollution which may have long term effect, may deprive the benefit of the present legislation. The Bhopal Mass Disaster which occurred in the year 1984 but still new claimants of compensation appear every day in view of the long range ramification of MIC. In view of this fact, it is suggested that the limitation of period may be subject to certain exception for example the 'long-term deseases'.

Now the time has come when Parliament should pass the Bill at the earliest to fulfil the longfelt need of an 'alternative forum' for administering just, fair and reasonable justice to the suffering and toiling masses.

CHAPTER 2

The Concept of Strict Liability

(A) CONCEPTUAL BACKGROUND

The rule in *Ryland* vs *Fletcher*[91] known as 'Rule of strict liability, had its origin in the law of private nuisance and has often been treated as a particular species of nuisance.

In order to better appreciate the concept it is necessary to go in detail with the *Ryland* vs *Fletcher* case. The facts of this case were as follows :

'B' a millowner, employed independent contractors, who were apparently competent, to construct a reservoir on his land to provide for his mill. There were old disused shafts under the site of reservoir. When the water was filled in the reservoir it burst through the shafts and flooded B's coal mines on the adjoining land. 'A' did not know of this nor was negligent in his act in this case 'A' was held liable for damages. The basis of liability was based on the classical exposition of Blackburn, J. who laid down.

> the person who for his own purposes brings on his lands and collects and keeps there any thing likely to do mischief if it escapes, must keep it in at his peril ; and, if he does not do so, is *prima facie* answerable for all the damage which is the natural consequence of its escape.[92]

Blackburn, J. during the course of his judgement limited the application of this doctrine when he observed :

> He can excuse himself by showing that the escape was owing to the plaintiff's default ; or that the excape was the consequence of his major or the act of God.

---

91. (1865) 3H SC 774 (Court of Exchequer) ; LRI Ex (Court of Exchequer Chamber) ; (1868). LRI 3 HL 330 (House of Lords).
92. (1866) LRI Ex. 265, 279-280.

In the present case none of the above exceptions was applicable and 'B' was held liable. The House of Lords, speaking through Lord Cairns L.C. held the defendant liable for "a non-natural use" of his land ; the learned Lord went to the extent to say that if something was brought on the land which was not naturally there and caused damage, the liability would be attracted.

Winfield preferred to call the instant liability as 'strict' rather than 'absolute' liability in view of various exceptions to this liability.[93] In view of this stand we have to be very clear in our mind about the distinction between the 'strict' and 'absolute' liability though at time confused stand is suggested.

The expression 'non-natural use' raises certain questions : What is natural and unnatural use ? Is 'non-natural use' the same as artificial use ? When will a use be termed as unnatural use ? Newark tried to define unnatural to mean an expression of the fact that the defendant has artificially introduec unto the land a new and dangerous substance[94] (*Emphasis supplied*). Whereas, Lord Moulton[95] and Lord Viscount Simon[96] had different meaning of the above expression, to them it meant.

> It must be some special use bringing with it increased danger to others and must not merely be the ordinary use of the land or such a use as is proper for the general benefit of the community.[97]

This meaning later on gained support. However, Lord Porter[98] in order to allow the principle to be appreciable for time to come, qualified the expression and laid down,

> What is non-natural should change in response to changing social conditions and needs.

In view of this, the expression 'non-natural use' has been given a limited meaning in the present time, provided the special use did not bring in 'increased danger.'[99]

---

93. Winfield, The Myth of Absolute Libaility (1926) 42 *LQR*, 37, 51.
94. Newark, Non-natural Users and *Ryland* vs *Fletcher* (1961) 24 *Modern Law Review*, 551, 561.
95. *Rickards* vs *Lothian* (1913) A.C. 263, 279.
96. *Read* vs *Lyons*, 1947, A.C. 156, See p. 169.
97. *Id* at 176.
98. *Winfield and Jolouricz on Tort* Ed. W.V.H. Rogers, 1989, 431-432.
99. *Balkrishnan Menon* vs *Subramanian*, AIR 1968, Ker. 151.

To sum up, the principle of strict liability will arise, if a person brings on his land and keeps there any dangerous thing which is likely to do mischief or cause an abnormal risk, if it escapes, he will be *prima facie* responsible for the damage caused by its escape even though the person concerned had not been negligent in keeping it there.

We have seen that the judiciary has diluted the rigid application of the principle of strict liability and now there is a demand in England[100] and also in the United States[101] to codify the strict liability principle so that protracted litigations do not held up its application and further, its application does not vary from case to case.

(B) IT'S APPLICATION IN ENVIRONMENTAL POLLUTION

In India the principle of *Ryland* vs *Fletcher* has received, over the years, support with the changing time. In *M.C. Mehta*, however, the Supreme Court did not think it necessary to follow the parameter of the English decision. The court rather opined that this rule of 'strict liability' was evolved in 19th century and it did not fully meet the needs of a modern industrial society with highly developed scientific knowledge and technology where hazardous or inherently dangerous industries were necessary to carry a part of the developmental programme. The court further pointed out in 19th century all these developments of science and technology had not taken place, and so law laid down at that time could not offered any guidance in evolving any standard of liability consistent with constitutional norms and needs of the present day economy and social structure.[102] The court was of the opinion that as new situation arose the law had to meet the new challenges and in this situation, the law could not afford to remain static. The Supreme Court, rightly, denounced adhearing to the out dated foreign legal order and opined :

> We have to evolve new principles and lay down new norms which would adequately deal with new problems which arise in

100. *Report of the Law Commission on Civil Liability for Dangerous Things and Activities*, No. 32, 1970, See also *Royal Commission on Civil Liability and Compensation for Personal Inquiry* Cmnd 7054, 1978, 1, Chapter 31.
101. Restatment 2nd ; s. 519.
102. *M.C. Mehta* vs *Union of India*, AIR 1987 SC 1086, 1098.

highly industrialised economy. We can not allow our judicial thinking to be constricted by reference to the law as it prevails in England or for the matter of that in any other foreign country. We no longer need the crutches of a foreign legal order.[103]

In the wave length of the Indianised jurisprudence of the 'strict liability', the court enunciated the rule as follows :

> Where an enterprise is engaged in hazardous or inherently dangerous activity and harm results to any one on account of an accident in operation of such hazardous or inherently dangerous activity resulting, for example, in escape of toxic gas, the enterprise is strictly and absolutely liable to compensate all those who are affected by the accident and such liability is not subject to any of the exceptions which operate *vis-a-vis* the tortious principle of strict liability under the rule in *Ryland* vs *Fletcher*.[104]

The basis of the new rule of strict, absolute and non-delegable liability were : Firstly, if an enterprise was permitted to carry on a hazardous or inherently dangerous activity for its profit, entails the condition that the enterprise engaged in such activity indemnify all those who suffered on account of such activity regardless of whether it was carried out carefully or not. Secondly, persons who were harmed as a result of such hazardous or inherently dangerous activity "would not be in a position to isolate the process of operation from the hazardous operations of substance or any other related element that caused the harms." The enterprise must be held strictly liable for causing such harm as a part of the social cost, for carrying on such activity, and lastly, the enterpise alone has the resource to discover and guard against the hazardous activity and to provide warning against potential hazard. Now coming to the distinction between the rule in *Ryland* vs *Fletcher* and the rule laid down in *M.C. Mehta*,

(*i*) The strict liability rule requires non-natural use of land and 'escape of inherently dangerous things. The absolute liability rule requires that the defendant should be

---

103. *Ibid.*
104. *M.C. Mehta* vs *Union of India*, AIR 1987 SC 1086, 1099.

engaged in hazardous or inherently dangerous activity, that the harm results to any one on account of accident.

(*ii*) The 'strict liability' will not cover the cases of persons within the premises. The 'absolute liability' makes no distinction between within premises or outside the premises.

(*iii*) The 'strict liability' is not absolute and is subject to many exceptions but 'absolute liability' rule is not only strict but absolute and is subject to no exception.

(*iv*) Another important point of distinction between the two rules is the award of damages. Damages awardable in 'strict liability' will be ordinary or compensatory but in absolute liability rule, court can allow exemplary damages and it further depends on the magnitude and capacity of the enterprise. The court in the *Mehta* case advocated that the larger and more prosperous is the enterprise, the greater must be the amount of compensation payable by it.

The rule of 'absolute liability' is an 'unique principle' but it has attracted criticisms. Chief Justice Rangnath Mishra in his judgement in *Union Carbide Corporation* vs *Union of India*[105], opined that judgement in *M.C. Mehta* was obiter. Upendra Baxi[106] joined this debate by saying that the whole discussion on *M.C. Mehta* in the *3rd October's* judgement was massive *obiter*. The *Mehta* principle was the foundation of the declaration of liability of the Union Carbide Corporation by Justice Seth of Madhya Pradesh High Court[107]. This judgement never came for consideration on full merit.

Prasant Bhusan, Counsel of one of the organisation Challenging the *Bhopal settlement*[108] emphatically said, "Now there is no such things as M.C. Mehta rule of absolute liability in India. It is finished." But the noted jurist Soli Sorabjee, did not agree with this contention and opined :

---

105. *Union Carbide Corporation* vs *Union of India* AIR 1992 SC 248.
106. In an Interview to magazine '*The Lawyer*' in Nov. 1991 issue.
107. *Union Carbide Corporation* vs *Union of India* AIR 1988 (NOC) 50 MP.
108. *Union Carbide Corporation* vs *Union of India*, AIR 1990 SC 273.

> As far as I know only Chief Justice Mishra has cast the doubt and frankly I think it was totally unnecessary and it is a complete obiter dicta ; the Mehta principle is still good law. An obiter dictim of one judge not assented by other can not change the effect of a binding precedent.[109]

Supporting the above arguments, he further contended,

> Mehta principle is not an unknown principle. As far as I know in some state courts and in thinking of American jurist there is convergence with the view taken in Mehta principle of absolute liability. The difficulty which the Chief Justice Rangnath Mishra envisaged was more imaginary than real.

The *Mehta* dictum has been further criticised by the foreign jurist as,

> Imposition of liability by judicial dicta regardless of general principles of law, which would 'undermine the credibility of Indian courts and lead to accusation of bias from foreign investor.[110]

However, the Supreme Court seems to have deliberately missed an opportunity to develop new principles in relation to multinational corporation operating with inherently dangerous technologies in the developing countries like India.[111] As the court itself said[112] that it would have examined various dimensions of this problem like the protection of environment, the possibility of ultra hazardous technology, standards of disaster, liability for multinational operating in developing countries, impact of exploitation of cheap labour, captive markets and the legal and constitutional safeguards against such exploitation. The court did not proceed to deal with these issues as the need for immediate relief to the victim of tragedy could not wait till these questions are elaborately examined and decided.[113]

---

109. In an Interview in November, 1991 Issue of Magazine "*The Lawyer*".
110. P.T. Muchlinski "The Right to Development and Industrialisation of less developed countries. The case of compensation for major industrial accidents involving foreign owned corporation, 1987, 50 M.L.R. 545.
111. The Bhopal case : Controlling ultra hazardous industrial activities undertaken by foreign investors. Muchlinski 1987, 50 M.L.R.
112. *Union Carbide Corporation* vs *Union of India* AIR 1990 SC 273, 283.
113. *Id* at 284.

It is, no doubt, true that the doctrine of absolute liability is best suited in case of environmental pollution and it answers the problem of 20th century's development process. But one should apply it with care and caution. There may be a case of unprecedented natural phenomenon resulting in damage. Will the absolute liability apply in such case ? Further, we have to consider its repercussion on India's economy. While India is opening its economy to foreign investor, can it afford to create a doubt in the minds of potential investors about the application of the principle of 'absolute liability' ?

The mass media in India has highlighted this apprehension when it records that several multinational corporation who were planning to invest in India imposed a precondition that the principle of absolute liability be not imposed on them.

To sum up, what is necessary is that 'the polluter must pay'. At the same time, we have to remember that the rigid application of the rule should not unreasonably retard the development process, otherwise it would be an end of all progress and development of a country. If this happens the country would breath its last breath. The rule should be what is called environmental friendly and not the enemy approach. And this brings us to what we may call a balanced approach in environment and development. The environmental lawyers[114] have advocated a statutory liability in this regard

> which will be attracted in case of an activity which contributes to, directly or indirectly to damage or injury to human beings or other elements of environment shall render any industry or class of industry or any enterprise involving hazardous or inherently dangerous activity responsible to strict and absolute liability.

It further made clear that such responsibility shall extend to :

(*a*) Compensate all human beings, and their property affected by any accident or any unforeseen happening, such damage or injury ;

---

114. *Environment Protection Act : An Agenda for Implementation* Pre. by Upendra Baxi, 987, 49-50.

(*b*) Comply with all rules or directions ... seeking to or ameliorate the damage or injury done to environment.

We have also to note that the judges while administering environmental justice while applying the statutory strict and absolute liability, must not allow their individual economic philosophy or remain emotionally attached with their viewpoint.[115]

---

115. See C.M. Jariwala, A Direction of Environmental Justice in India : A Critical Appraisal of the 1987 case law. (Unpublished) wherein the author dealt with the approach of individual judges of the Supreme Court of India and the High Courts.

CHAPTER 3

Compensation : A Remedial Measure

(A) DEVELOPMENT OF COMPENSATION AS A REMEDIAL MEASURE

Article 32(1) of the Indian Constitution provides for the right to move the Supreme Court by appropriate proceedings for the enforcement of the fundamental rights. The Supreme Court under article 32(1) is free to devise any procedure for the enforcement of fundamental right and it has the power to issue any process necessary in a given case. In view of this constitutional provision the Supreme Court may even give remedial assistance which may include compensation in "appropriate cases."

The year 1980 is an important period when the Supreme Court of India adopted a dynamic approach in the field of processual justice to which Prof. M.P. Jain named as an era of "manifestation of the dynamic constitutional jurisprudence."[116] Such manifestation was first visible in *Khatri* vs *State of Bihar*[117], where the question was "the exploration of new dimension" of the right to personal liberty guaranteed by article 21. Bhagwati J. answered this question in the following words :

> ....why should the court not be prepared to forge new tools and devise new remedies for the purpose of vindicating the most precious of the precious fundamental right to life and personal liberty.[118]

In a trend setting judgement in *Rudal Shah* vs *State of Bihar*[119], the petitioner in a habeas corpus writ petition prayed the Supreme

---

116. M.P. Jain *'Indian Constitutional Law* ; 1987, *Id* at 599.
117. AIR 1981 SC 928.
118. AIR 1981 SC 928, 930.
119. AIR 1983 SC 1086.

Court to pass an appropriate order for the payment of compensation for his illegal detention for fourteen years.

The supreme Court, in view of injustice with the petitioner, ordered the State of Bihar to pay the petitioner a sum of Rs. 35,000. The court felt that if it refused to pass an order of compensation in favour of petitioner, according to the court, "it will be doing merely lip service to the fundamental right to liberty which the State Government has so grossly violated."[120] This was only an interim measure and according to the court it did not preclude the petitioner from suing the State and its officers for appropriate damages.

The Supreme Court formulated a general guideline to award compensation to the victim of State violence which included :

(*i*) Article 21 which guarantees the right to life and liberty will be denuded of its significant content if the power of the Supreme Court is limited to passing orders of release.

(*ii*) The violation of right guaranteed under article 21 can reasonably be prevented and due compliance with the mandate of article 21 secured, is to mulct the violators in the payment of monetary compensation.

(*iii*) Administrative sclerosis leading to flagrant infringement of fundamental right can be corrected by other methods open to the judiciary.

(*iv*) The right to compensation is some palliative for the unlawful acts of instrumentalities which act in the name of public interest and which present for their protection the power of the State as Sheild.

(*v*) The respect for rights of individual is true bastion of democracy. Therefore, state must repair the damage done by its officer to the petitioner.

In *Sebastin M. Hongray* vs *Union of India*[121] two persons were taken to the military camp by jawans of the army. Their wives filed a writ of habeas corpus in the Supreme Court, asking the Government

120. AIR 1983 SC 1086, 1089.
121. AIR 1984 SC 1026.

of India to produce aforesaid persons before it. The government failed to produce them. The truth was that these persons had met an unnatural death. In the circumstances the Supreme Court keeping in view the torture, agony and mental oppression through which the wives of the person in question had to pass, required that as a measure of exemplary costs the government must pay Rs. one lakh to each of aforesaid women.

In *Boma Charan Oraon*[122] an under trial prisoner was detained in a lunatic asylum for six years after he had been certified as fit for discharge from the detention. The court awarded him Rs. 15,000 as compensation. The court observed that no amount could possible compensate a sane person for living in a lunatic asylum for six years.

In *Bhim Singh* vs *State of J.K.*[123] the Supreme Court awarded a sum of Rs. 50,000 to the petitioner as compensation for the violation of his constitutional right of personal liberty under article 21 of the Constitution. The petitioner, a member of Legislative assembly of Jammu and Kashmir was arrested and detained in police custody and deliberately prevented from attending the session of Legislative Assemble. The police officer acted deliberately and malafide and judicial officer helped them in this process. The Supreme Court applying the principle evolved in the *Rudal Shah*[124], awarded monetary compensation to the petitioner. Following this decision, in *SAHELI* vs *Commissioner of Police*.[125] The Supreme Court directed the government to pay Rs. 75,000 as compensation to the victim a boy of 9 years who died because of beating by police officer.

Thus, Indian judiciary has made unique contribution in the field of human right by forging new remedy to compensate the poor people against the mighty power of the 'State'.

(B) IT'S ROLE IN ENVIRONMENTAL POLLUTION

It was in the third *M.C. Mehta case*[126] the Supreme Court addressed itself to the principle of liability of industries, engaged in

---

122. *B.C. Oraon* vs *State of Bihar*, Decided on 12th August, 1983 See *Hindustan Times*, Delhi, August 13th 1983.
123. AIR 1986 SC 494.
124. AIR 1983 SC 1086. See also *A.S. Mittal* vs *State of U.P.* AIR 1989 SC 1070.
125. AIR 1990 SC 513.
126. AIR 1987 SC 1086.

inherently dangerous and hazardous activities to compensate the victims of the accident. The Supreme Court reiterated its stand that apart from issuing directions it can under article 32, forge new remedies and fashion new strategies designed to enforce the fundamental right.[127] The court went on to say[128] that the power under article 32 was not confined to preventive measure when fundamental right was threatened but it extended to the remedial measure when the rights were violated. A contrary position the court observed, would robe article 32 of all the efficacy and render it impotent and futile.[129] The Supreme Court could award compensation in writ petition itself. The court further pointed out, we cannot adopt a hyper technical approach which would defeat the end of justice.[130] The Supreme Court put emphasis on the word appropriate cases. It observed :

> We are deliberately using the words in 'appropriate cases' because we must make it clear that it is not in every case where there is breach of fundamental right committed by violator that compensation would be awarded by the court in a petition under article 32.[131]

What are the appropriate cases, the court left the question open. However, the compensation cases referred above give a clear indication that it should affect a large number of persons and in particular the poor. The significance lies in the court's formulation of the principle of measure of liability in case of industrial hazard.

The court laid down that the measure of compensation must be correlated to the magnitude and capacity of the enterprise so that the compensation could have a deterrent effect.[132] The court directed the Delhi Legal Aid and Advice Board to take up the cases of all these who had suffered on account of gas leakage to file action on their behalf in appropriate court for claiming compensation

---

127. *Id* at 1089.
128. *Id* at 1091.
129. *Ibid.*
130. *M.C. Mehta* vs *Union of India*, AIR 1987 SC 1086, 1089.
131. *Id* at 1091.
132. *M.C. Mehta* vs *Union of India*, AIR 1987 SC 1086, 1099.

against Sriram. In the instant case[133], the court did not award compensation in the writ proceeding, however, it laid down the principle of liability and left the issue of the amount of compensation in each case to be decided by the appropriate court. The Supreme Court also directed the Delhi administration to provide necessary fund to Delhi Legal Aid and Advice Board for the purpose of filing and prosecuting such actions. It may be recalled that in the second *M.C. Mehta*[134] case Supreme Court modified the condition relating to payment of compensation. The Supreme Court held that it was the individual responsibility of the Chairman, Managing Director and officer to compensate the victims of gas leak. However, in case of officer who was acting as Head of caustic chlorine plant, the liability of such officer was limited to the extent of his annual salary with allowances.

It may be mentioned that in the whole compensation exercise in the *M.C. Mehta* case, the victims of the oleum leak were the only beneficiaries. The would be victims, the animals and plants who may be or were affected were left out. The oleum gas leak has not only immediate but also long term effect. This part has been completely forgotten in the entire exercise. It may be further pointed out that the judicial verdict handed down almost six years ago has yet to reach to the victims. The same is the fate of the Bhopal disaster. This judicial inaction has raised a question on the role of the judiciary. Is not the time right that the court must not simply deliver the judgement but also see that the beneficiaries are benefited ?

Aftermath of the M.C. Mehta

The principles formulated in *M.C. Mehta* have been applied by the Supreme Court in the *Union Carbide* case. Just after the *M.C. Mehta* case the Supreme Court was once again confronted with the question of ordering compensatory remedy in a mass disaster.[135]

The Bhopal District Court followed *M.C. Mehta* principle and passed an order against *Union Carbide Corporation* for depositing

---

133. *M.C. Mehta* vs *Union of India*, AIR 1987 SC 965.
134. AIR 1987 SC 982.
135. AIR 1990 SC 273.

Rs. 350 crore as interim compensation. On appeal the Madhya Pradesh High Court[136] reduced the compensation to Rs. 250 crore. When the matter came up before the Supreme Court, it passed a final order of settlement comprising all the claims civil and criminal and quantified the amount to 470 million dollar.[137] This settlement order of 470 million dollar was again challenged before the constitutional bench of Supreme Court, on October 3, 1991[138] which unanimously upheld the settlement and directed the Central Government to make good deficiency if the settlement fund was found to be insufficient.

The judicial dynamism in the *M.C. Mehta* activated Parliament to impose statutory liability on those dealing in hazardous substance. The Public Liability Insurance Act, 1991, was passed to provide relief to the members of the general public who become victims of industrial accidents. This legislation expressly excludes the workmen from its purview since they are entitled to relief under the workmen's compensation Act, 1923. For the purpose of fixing liability for payment of compensation under the Public Liability Insurance Act, 1991, no distinction has been made between public sector and private sector or between an individual or government official. National as well as multinational are equally liable. Any person who has control over handling of any hazardous substance would also be liable to pay compensation under the Act.

The rate of compensation prescribed under the present legislation is given in the following table.[139]

---

136. AIR 1988 (NOC) 50 (MP). For interesting literature on the *Bhopal case*, see also
   (*a*) *Inconvenient Forum and Convenient Catastrophe*. The *Bhopal case*, 1986 prepared by Upendra Baxi.
   (*b*) *Mass Disaster and Multinational Liability. The Bhopal case*, 1986, prepared by Upendra Baxi and Thomas Paul.
   (*c*) *Valiant Victims and Lethal Litigation. The Bhopal case*, 1990, prepared by Upendra Baxi and Amita Dhanda.
137. AIR 1990 SC 273.
138. AIR 1192 SC 248.
139. For critical discussion of Public Liability Insurance Act, 1991, See Public Liability Insurance Act, 1991 *Scope for making provision more effective* by S.N. Singh, 5 *Corporate Law Adviser* May (1991). See also Mehdi, Ali, *Public Liability Insurance Act, 1991—A Critical* Appraisal, AIR 1992 Journal 36.

TABLE

S.No.	Nature of injury/damage head	Amount
1.	Reimbursement of medical expenses	Maximum Rs. 12,500
2.	Fatal accidents	Rs. 25,000 per persons + reimbursement of medical expenses as above
3.	Permanent/partial/disability/ other injury or sickness	Reimbursement of medical expenses as above + cash relief based on percentage of disablement certified
4.	Total permanent disability	Rs. 25,000 + reimbursement of medical expenses
5.	Loss of wages due to temporary partial disability necessitating minimum three days hospitalisation- and victim is above 16 years age	Fixed monthly relief with maximum of Rs. 1,000 per month for period not exceeding three months
6.	Damage to property	Actual damage or Rs. 6,000 whichever is lower

It is interesting to note that Parliament, in the concern for the victims of an accident, amended the Act of 1991 in the year 1992 and provided for the establishment of Environment Relief Fund for which the owner shall contribute an amount equal to the premium money. Initially this amount will be deposited with the insurer who shall remit the money to the appropriate authorities within the prescribed period. The collector shall arrange to pay from the fund to the affected persons. The present Amendment Act, 1992 lays down the maximum amount of policy not exceeding five crore rupees which the owner of any undertaking handling any hazardous substance. The Act, in order to compensate the victims to a reasonable extent, requires that even after the payments from the insurer and the Environmental Relief Fund, still the victims are not adequately compensated, the remaining viability shall shift on the owner of the hazardous undertaking. It may be pointed out that in this compensatory exercises, the State's involvement in compensation is left out. In the present case it was because of the State's inaction that Sriram grew in such a thickly populated area. Is not the State responsible in this episode ?

The aftermath of the *M.C. Mehta* case is encouraging. The compensatory remedy is gaining a quantified statutory base. This innovation has yet to find a place in other jurisprudence. The content of the compensation table may not be acceptable in dollars, pounds and even in the Indian currency but it is a starting point to specifically quantify degree of compensation in varying injuries.

PART V : HAZARDOUS INDUSTRIES AND LEGAL CONTROL

Hazardous Industries and Legal Control

(A) DEFINITION OF HAZARDOUS INDUSTRIES

'Hazardous Industry' has not been expressly defined in any environmental legislation. However, section 2(b) of the Environmental Protection Act, 1986 defines environmental pollutant and section 2(e) defines hazardous substance.

Section 2(b) defines an environmental pollutant as "any solid or gaseous substance" injurious or tending to be so, to environment. Section 2(e) defines hazardous substance as meaning any "substance or preparation which by its reason of its chemical or physio-chemical properties is liable to cause harm to human beings, other living creatures, plants, micro-organisms, property or the environment."

Both the above sections indicate that 'pollutant' and 'hazardous substance' are two different concepts for the environmental law. It is not necessary that all pollutants are necessarily 'hazardous' substances' but all hazardous substances are pollutants. The key idea underlying and also be some extent explicit, in both these definition is probability of harm or injury to the environment as defined under the Act.[1]

A substance, a process, a preparation can be an environmental pollutant or hazardous substance. The reference to 'solid, liquid or gaseous properties for pollutants and to 'chemical or physio-chemical properties' for 'hazardous substance' creates potential for legal and scientific (expert evidence) quibbling in judicial fora.[2]

In view of the above distinct approach, it is suggested that the handling of the above substances, a different approach is necessary.

1. *Environment Protection Act : An Agenda for Implementation*, prepared by Upendra Baxi,1987, 6.
2. *Id* at 6.

The term 'hazardous substance' has been defined in section 2(d) of the Public Liability Insurance Act, 1991, to mean "any substance or preparation which is defined as hazardous substance under the environmental protection Act and exceeding such quantity as may be specified by notification, by Central Government."

The definition under the Environment Protection Act, 1986 will not apply to Public Liability Insurance Act, 1991, unless the substance in question exceeds in such quantity as specified by the Central Government.

When a substance is considered hazardous for the purpose of one legislation dealing with environment, it does not appeal to reason why that substance should not be treated hazardous for another legislation on the same subject, more particularly since the Public Liability Insurance Act, 1991 is a beneficial legislation to protect the members of general public who are killed or injured because of handling of hazardous substance by another who is exploiting it for personal benefit.[3]

The Government of India framed the Hazardous Wastes (Management and Handling) Rules, 1989, in the schedule to the rules, makes categories of waste, for example in waste category number one comes so cyanide waste, in number two metal finishing wastes is given place, waste containing water soluble chemical, compounds of lead, copper, zinc, chromium, nickle is placed in the third category. The schedule has given 18 wastes categories.

However, Section 2(k) of the Air (Prevention and Control of Pollution), Act, 1981, defines industrial plant means any plant used for any industrial or trade purposes and emitting any air pollutant in the air. Section 2(a) of the said Act defines Air Pollutant means any solid, liquid or gaseous substance present in the air in such concentration as may be or tend to injurious to human beings or other living creatures or plants or property or environment.

The hazardous chemicals, dangerous goods are divided into class and division by United Nation Committee of experts on the transport of dangerous goods. This list will provide us a fair idea about nature of 'dangerous goods'. The hazardous industries' play

---

3. Dr. S.N. Singh; Public Liability Insurance Act, 1991; Scope for making provisions effective. 5 *Corporate Law Advisor*, May 1991, 234.

an important role in manufacturing, processing and packaging of such 'dangerous goods'. The list divides the dangerous goods into following classes and divisions:

Class I	:	Explosive
Division I	:	Substance and articles which have a mass explosion hazard,
Division II	:	Substance and articles which have projection hazard but not a mass explosion hazard
Division III		Substance and articles which have fire hazard or a minor projection hazard or both but not a mass explosion hazard
Division IV		Substance and articles which present no significant hazard
Division V	:	Very insensitive substance which have mass explosion hazard;
Class 2		Gases, compressed, liquified, dissolve under pressure or deeply refrigerated
Class 3	:	Inflammable liquids
Class 4	:	Inflammable solids, substance liable to spontaneous combustion; substances which, on contact with water, unit inflammable gases;
Division 4.1	:	Inflammable solids,
Division 4.2	:	Substance liable to spontaneous combustion
Division 4.3		Substance which on contact with water, emit inflammable gases.
Class 5	:	Oxidising substance, organic peroxides
Division 5.1	:	Oxidising substances
Division 5.2	:	Organic peroxides
Class 6	:	Poisonous (toxic) and infectious substances
Division 6.1	:	Poisonous (toxic) substance
Division 6.2	:	Infectious substances

Class 7 : Radio active substances

Class 8 : Corrosive

Class 9 : Miscellaneous dangerous substances[4]

It appears from the above discussion that these industries which are using hazardous substance in manufacturing process or using hazardous substances for any other purpose are to be treated as 'hazardous Industries'.

It is interesting to note that in 1984 the estimated annual hazardous waste generation in India has about 0.3 million tonnes per year and it is estimated that now it may be one million tonnes per annum.[5] The above data show that the government has failed to handle the situation and it is time that we must take the stock of the environment unfriendly industrial approach and set the hazardous homes right.

(B) EMERGING WESTERN VIEWPOINTS

At present, in western countries three fold environmental principles are in the operation to protect and preserve the environment.[6]

1. Preventive principle;
2. The precautionary principle;
3. The polluter pays principle.

1. The Preventive Principle

The creation of pollution should be prevented at its source. This principle is self, evident in second European Community Action Programme (1987-1991) which talks of 'preventing the creation of pollution at its source. Complete prevention of pollution is both practically impossible and politically infeasible. It is argued that complete prevention of pollution will derail us from the path of growth and development. So the principle is really no more than goal or objective.

---

4. See *Environment Protection Act : An Agenda for Implementation*, prepared by Upendra Baxi, 1987, 62.
5. N. Sarkar, Industry alone not to blame, *The Hindu Survey of the Environment*, 1992, 137.
6. Integrated Pollution Control in Environment Protection Act, 1990: A Coming Age of Environmental Law by Michael Purdue; *Modern Law Review*, July 1991, 535-536.

2. The Precautionary Principle

This principle seems to have originated in West Germany. The German word 'Vosorgprizip literally means precaution or *foresight.*[7] However, it has come to mean the minimisation of the risk of pollution and it is interesting to note that in Britain, in recent white paper, This Common Inheritance: Britain's environmental strategy, adopts a qualified version of this principle when its states that "where there are significant risk of damage to the environment, the government will be prepared to take precautionary action to limit the use of potentially dangerous pollutants, even where scientific knowledge is not conclusive, if the balance of likely cost and benefit justifies it.[8]

This last reference to balance emphasises again that this is not absolute principle and its application must be guided by proportionately principle United States Clean Air Act, 1982 is a statutory adoption of this principle.

3. The Polluters Pays Principle

The basic idea behind this principle is that costs should be internalised by the producer, of the pollution and not passed to some one else. What makes this a very illustrative principle is that it is not clear whether it means the person producing pollution should pay all the costs of eliminating or whether he must pay to continue pollution or even that there is acceptable level of pollution and if the polluter does not exceed that level, he does not have to pay at all. This phrase is also misleading because in practice the polluter indirectly shifts this burden on consumer through high prices. This can not be justified though polluter may think it proper to share the cost with consumer. All these three principle can be traced in equally attractive but vague terms 'sustainable development' which argues that fairness to future generation *entails cautious* anticipatory policies which out low short term gains.[9]

---

7. Twelfth Report of the Royal Commission on environmental Pollution cm 310 (February 1988). Cited in 54 *Modern Law* Review, 1990, 535.
8. Cm 1200 (London: HMSO: 1990 *Id* at para 1.18.
9. For an exhaustive analysis of this term see pearce, *Blueprint for Green Economy*, 1989, Chap.2.

Need for Environmental Impact Assessment

Environmental impact assessment (EIA) has emerged as the most basic environmental law establishing (ELA) and its procedures has its origin in the National Environmental Policy Act (NEPA) in the United States.

An environmental assessment is made initially to determine if the action's effect may pose significant adverse consequences such as degrading an ecosystem, and if so, agency must prepare an environment impact assessment state (EIS) describing effects and alternative action possible to avoid the adverse effects.[10]

Nearly two decades of employing EIA federal procedure in United States have established its value.[11]

The environment impact assessment has also been adopted by many other nation and states. Out of this practice a standard version of EIA has emerged.[12]

While India is facing catastrophe like 'Bhopal gas tragedy' Indian policy makers have deliberately ignored the dynamic revolution that has taken place in field of environmental law all over the world; by simply stating that it will be better for existing agencies to implement the law.

Even Indian policy makers have not bothered to enforce dormant provision of existing law. Section 3 of Environmental (Protection) Act, 1986 provides for creation of 'Environmental Protection Authority' similar to the agencies operating in developed countries to assess the impact of pollution on environment. The primary function of Environmental Protection Authority is post project (including hazardous project) analysis as well as impact assessment prior to project. This post project review is extremely valuable as check on the quality of EIA process and assess over time the effectiveness of measure used to mitigate or avoid adverse impact of project.[13]

---

10. N.A. Robinson, 1982: SEGURA's Sibbings precedents from little NEPA in the Sister states, *Albany Law Review* 46, 1155.
11. Impact *Statements Administrative Reform*, 1984, Standford Press, California.
12. See Figure 1 at 150.
13. De Soet, Spaink and Meijers, 1987, Post project analysis as required by the EIA Act. In the Netherlands UNICEF Seminar on Environmental Impact Assessment (Warsaw).

The United Nation Environment Programme Group of experts on environmental law in January, 1987 urged the use of EIA especially in case of possible transfrontier impacts.

Environment impact assessment depends for its effective operation on following conditions:

(*a*) The availability of adequately trained scientists to ecological and other relevant data.

(*b*) The funds to pay for analysis.

In most instances expenses of EIA are deemed as cost of action under analysis where a project cannot pay the cost of EIA the government budgetry procedure must allocate funds to do so.

Thus it can be forcefully urged that Environment Impact Assessment is a must to protect Indian public and environment from possible catastrophe like 'Bhopal gas tragedy' and Sriram's oleum gas leak case.[14]

The Priorities of Environmental Law and Policy in 21st Century

Environmental law is likely to develop further in coming years. The Swedish Academy of Sciences has identified ten key environmental issues with which environmental law specialists, scientists and government officials would have to cope.[15] French Ministry of Environment and Ministry of Urban Planning and Housing has prepared an independent survey of similar issues.[16]

The management of hazardous chemical, *processes* and waste has been accorded top priority among ten key Swedes environmental issues. Similarly in French survey toxic pollutants, occupies sixth place among ten key environmental issues.

---

14. Environmental law experts have also recognised the necessity of such assessment. For detail see 'Environment Impact Assessment Group': *Environment Protection Act, An agenda for Implementation*, prepared by Upendra Baxi, 1987, 9. See also P.S. Sangal, Some Recent Developments in the field of environment., Prospect for the Future, 13, *Delhi Law Review*, 1991, 44-71.

15. Conference on Environmental Research and Management Priorities for the 1980's. 1983 *AMBIO* 12 (2) : 58 Stockholm.

16. Consultation prospective sur l'environment 1982. Minister de 1, environment et minister de 1,"Urbanisme et du logement Min. de l' Env't Paris.

Thus it can be said that the major emphasis will continue to be placed on protecting the public health and natural environment from industrial waste on the basis of studies of 'priority' of these two most 'developed countries'. Chemical waste with toxic or hazardous effect will attract major attention, with prime focus on reducing the volume of such wastes by regulating them or process change to generate lesser volume.[17]

The latest theory which is currently gaining popularity in developed countries is RISK ASSESSMENT THEORY.

Risk Assessment Theory

Since every government has financial constraints, it can not cope equally with all environmental issues due to limited availability of fund.

Risk Assessment is the process by which we estimate the *probability* that a particular polluting substance will cause human or environmental damage.

Risk Assessment is a technique to weigh the more serious of risks, and rank them accordingly. At present, the attention of most developed countries seems to be devoted to the new concept of risk assessment, and as the consensus develops, the technique of risk assessment will be added to environmental law.[18]

Once the risk assessment is done, it can be used for risk management decisions, the requirement of relevant legislation, the effect of decision on other social values, the degree and distribution of current risks and proposed risk reduction and the level of confidence in the risk assessment.

The strategy based on risk reduction requires major improvements in two area.[19]

1. The first is communication, and
2. The second relates to increasing our understanding of real risk and distribution.

Thus it may be said that changing policy emphasis from

---

17. See Seminar report for detail discussion. Symptom rehabilitation "The Environmental Law Framework", Prof. N.A. Robinson, International Symposium in Banaras Hindu University, 11-16, December, 1987, 24.
18. *Quantitative risk assessment in regulation*, Ed. L. Lave, 1982.
19. Environmental quality 1985, U.S.A., 16th *Annual Report* pp. 22-23 of Council on Environmental Quality.

pollution control to reduction of risk requires new form to achieve the object. Technology based command and control regulations is not useful in dealing with hazardous industries. The alternative is to adopt policies that will turn the interest of polluters against polluting, incentive based on risk reduction are worth trying on significant scale. States[20] have used fee system for hazardous waste disposal, constructed them as to favour the safest form of disposal and penalise the production of hazardous waste.[21]

We may sum up by saying that Indian legislature and executive should recognise the latest development in field of environmental law and should adopt measures to upgrade the existing laws to suit the needs of 21st century. Otherwise, we would be buying short term survival for a long term suicide. Secondly, an uniform and integrated approach is necessary to deal effectively with the problem of environmental pollution.[22]

(C) LEGAL CONTROL IN INDIA

(i) **Retrospect**

According to Hobbes, man is by nature nasty, brutish and violent and fear or sanction which is inherent in law is necessary to maintain order in society. This perception that something needs to be controlled provide impetus for legislation which becomes primary factor in regulation and control. The emergence of regulatory policy relating to hazardous industries owes its origin to this philosophy. Justice Krishna Iyer's observation reflects the similar perception. His observation merits our attention:

> The unconscionable industrialisation, the unpardonable deforestation and inhuman extermination of living species betry an exploitative brutality and anti-social appetite for profit and pleasure incompatible with humanism and conservationism.[23]

---

20. Such Act for the first time in India was passed in Indian Parliament on December 1, 1991 Water (Prevention and Control of Pollution) Cess (Amendment) Bill, 1991.
21. For detail discussion see Environmental Quality, 1985,
22. See for an interesting comparative environmental law treatment including Indian law in N.A. Robinson, Marshulling Environmental Law to resolve the Himalaya-Ganges Problem, 13 *Delhi Law Review*, 1991, 1, 24.
23. V.R. Krishna Iyer; *Justice to nature, Environmental* Pollution and the Law, 1984, 95.

This observation explicitly explaines the necessity of the regulation of the hazardous industries.

FIGURE 1*

(N.A. ROBINSON)

ENVIRONMENT IMPACT ASSESSMENT PROCESS (EIA)

1. Agency Action.

2. Notice to affected or interested authorities to agree on lead agency

3. Determination whether action will significantly affect the environment

4. If positive, proceed to Scope EIA

   4A. If negative proceeded to action

5. Study and prepare draft, written, environmental impact assessment report

6. Disseminate draft to other agencies and public for comment

7. Respond to comment and prepare final report

8. Decide on acceptable impact and modifying action to mitigate adverse effects

9. Undertake action

10. Post action audit to ascertain effectiveness of modification and assessment

Developing countries including India, have not yet well developed environmental law to cope with emerging problem of unplanned industrialisation and urbanization. In contrast, the legal system of western countries have environmental laws to certain

* Prof. N.A. Robinson, Director, Centre for Environmental Legal Studies, Prec University School of Law, White Plains NY, USA.

extent effectively control the environmental pollution. Professor Wali and Burgess observe:[24]

> On balance, however, the United States and many other nations are light years ahead of where they were in the days of their grand parents.

Now coming to regulation and control of hazardous industries; we will examine existing constitutional provisions, legislative provisions and provisions of environmental laws which are in operation.

1. **Constitutional Position**

Indian Constitution does not provide any specific place to environment or environmental pollution in the distribution of legislative power under the Seventh Schedule. However, the provisions could be impliedly gathered from articles 249, 250, 252, and 253 of the Indian Constitution. For instance, Union Government has enacted Water (Prevention and Control of Pollution) Act, 1974 under article 252, to check and control pollution of water and the Acts of 1981 and 1986 had the source of power under article 253, Entries 52 and 7 deal of list I with power of Parliament with respect to industries, the control of which is declared by Parliament by law to be expedient in the public interest and industries declared by law to be necessary for defence purpose. Apart from these two types of industries, the state legislature has the exclusive legislative power over the rest of industries under entry 24, list II. It is obvious from the examination of the above lists that hazardous industries may be controlled and regulated under list I entries 7 and 52 and also under entry 24 in list II.

The Constitution (forty second amendment) Act, 1976 brought in a new changes in the field of environmental law but it left out the source of legislative power untouched. The time has come[25] when the constituent power must fill this important gap to avoid any uncertainty. The solution lies in a cooperative approach

---

24. Wali and Burgess; The interface of ecology and law: Science, the legal obligation and public policy, *Syracuse Journal of International Law and Commerce*,1985, 12.221, Syracuse, N.Y.
25. See for repeated demands in this regards, S.L. Agrawal (Ed) *Legal Control of Environmental Pollution*, 1980, 215-217; Paras Diwan (Ed.) Environment Protection, 1987, 14, 135.

and this object can be achieved by a concurrent jurisdiction of Parliament and State legislature over the subject of environmental pollution.[26]

2. Legislative Padyatra

There were occasions when the India" legislature attempted to minimise the impact of hazardous industries through certain legislations. These legislations were enacted to control unplanned growth of hazardous industries. For example, Atomic Energy Act, 1962, Inflammable Substance Act, 1952, Explosive Act, 1952, Insecticide Act, 1968 and Mines Act, 1952. In 1951, the Industrial (Development and Regulation) Act, 1951 was passed. Section 15 of this Act authorised the Central Government to order investigation against those industries which manage their affairs in manner highly detrimental to public interest. Under Section 18-A Government was authorised to assume management and control of an industrial undertaking engaged in scheduled industry. In certain emergent cases government was further empowered to takeover certain industries even without any investigation. Industrial Policy Resolution, 1948 empowered the Central Government to make planning and regulation in national interest. Heavy chemical and fertilizer were included among 18 basic industries. Thus, it is evident that hazardous industries were mainly under the control and supervision of the Central Government.

(*i*) *Location and Establishment of Hazardous Industries* : According to Air Act, 1981, previous consent of State Board is necessary to operate any industrial plant in air pollution control area.[27] The State Government is empowered to declare any area within State as pollution control area.[28] Such consent[29] is given subject to the control equipment of such specification as the State Board may approve in this behalf, shall be installed and operated in the premises where the industry is carried on or proposed to be carried on.

---

26. C.M. Jariwala, Changing Dimension of Indian Environmental Law, ed. P. Leela Krishnan, *Law and Environment*, 1992, 13.
27. The Air Act, 1981, Section 21(1).
28. *Id* Section 19.
29. *Id* Section 21(5) (i) to (iv).

(*ii*) *Monitoring and Controlling Agencies* : The Central Board and the State Board are mainly responsible to control and regulate the pollution caused by hazardous industries. At present one Central Board and 21 State Boards are functioning.[30] The Central and State Water Boards have also to act as respective Central and State Air Boards which have been renamed in 1987 as the Central or State Pollution Boards. The pollution boards have not only to monitor and control pollution but also they have to constantly informed Parliament about the development with respect to pollution of water and air. The Environment (Protection) Amendment Rules 1992 requires the central pollution board to submit an annual report on the table of Parliament on the guidelines laid down therein. There are environmental protection council in 14 States and the three Union Territories, which have helped in curbing pollution caused by hazardous industries.

Minimal National Standard (MINAS) for pollution discharge from specific industries have been formulated and control measures are being implemented in phased manner. It is the duty of the person carrying on hazardous industry not to allow emission of pollutant in the excess of standard laid down by the board. It is also the responsibility of the person concerned to furnish information to the board[31] regarding emission of excess pollutant in the atmosphere due to accident or otherwise.[32] Further, no person shall knowingly cause or permit any poisonous, noxious or polluting matter into any stream or well.[33] There is duty to inform board if such discharge take place due to accident or any other reason.[34]

Every person carrying on any industry or process or handling hazardous substance shall be bound to render all assistance to the person empowered by Central Government, if he fails to do so without any reasonable cause or excuse, he shall be guilty of an offence.[35]

---

30. Nagaland and Arunachal Pradesh are only States where these boards are not functioning.
31. Air Act, 1981, Section 22.
32. *Id* Section 23.
33. Water Act, 1974, Section 25(i).
34. Air Act, 1981, Section 22.
35. Environment (Protection) Act, 1986, Section (2).

An Appraisal of Function of Monitoring and Controlling Agencies

1. Actually, both Water Act, 1974 and Air Act, 1981 prescribe no qualification and experience for appointment as member in State Board. State Boards mostly consists of government nominated members. Many of the chairmen of such boards, are political appointees, often prone to undue influence by industrial interest to turn blind eyes to violators of factory and other laws.

The official Annual Report of 1988-89 gives on ideas about enforcement of these regulations. It says that since the enactment of water and air Acts there were total of 3251 prosecutions. Under the Act of 1986, the Central Government initiated action against 82 units for not following necessary pollution control measures.[36] It is ironical that only 35 cases filed before these boards have been settled speaks for itself.[37]

The public sector projects which have been in existence for 15 years or so are among the worst offenders. The olds unit tannery at Kanpur, Calcutta, Madras have also not been able to set up their houses in order, and the government had to work out a plan for the common effluent treatment plant for tanneries.

The Power Ministry has drawn up a list of 189 power units that are complying with Central Pollution Control Board Standard, 79 are partially complying and 75 units have not even a programme for the pollution control. Similarly the Food Ministry has reported that 232 sugar factories have installed effluent treatment plants and 190 have installed flue gas treatment units.

2. The consent of the boards for new or altered outlet or for new discharge of pollution have become meaningless, because these acts itself provides that such consent shall be deemed to have been unconditionally given on expiry of period of four months from the date of making such application unless the consent is given or refused earlier.[38]

3. The power given to board are not adequate and whatever power they have are more of advisory or investigatory in nature.

---

36. The Annual Report 1988-89, p. 46.
37. Upto 3rd December, 1991 ; Reported in *Times of India*, 3, 1991.
38. See Section 25(7) of Water Act, 1974, and Air Act, 1981.

4. Boards have not been provided with enforcement power to take coercive or punitive measures against polluters. It is curious to note that boards can take action if some trade effluent is found in stream or well but they have no power to prevent apprehended pollution. For this purpose, boards are required to move the court for obtaining orders.

5. Hazardous industries are not cooperating with pollution control boards in combating the pollution. Hazardous industries did not think it necessary to apply for consent order from such boards. Commenting on this state of affairs, justice Krishna Iyer says :

> The corporate capacity to outwit the law is not peculiar to pollution control measure but to all economic regulations. This may be called blue chip syndrome in American diction.[39]

6. It has been found that industries did not obey the conditions imposed by the boards regarding safety measures on very flimsy ground such as absence of technology for treatment plant, non-availability of land and absence of proper disposal points.

However, some industries installed anti-pollution devices but never put them into operation to save the expenses. Even the board in its present position could monitor the plant once in a year. Several industries manipulate the provisions of law one way or other.[40]

7. It has been noticed that multiplicity of controlling agencies often retard the process of effective monitoring and control of pollution. The controlling agencies at Central, State, regional and municipal level often work at cross purpose. Instead of combating the problem of pollution, the agencies concerned entangled themselves in legal niceties and waste much of time to overcome them.

The biggest enemy of environment is unholy nexus between corrupt officials of board, politicians and polluting industries. N. Gunnigham rightly pointed out[41] :

---

39. V.R. Krishna Iyer ; *Environmental Pollution and the Law, 1984, 13.*
40. See for more detail Shyama Bhardwaj. *Management of Environment Pollution in India ; The New Environmental Age ;* Ed. R.K. Sapru and Shyama Bhardwaj, 1990, 65.
41. N. Gunnigham ; Power, Politics and Environment ; *13 Banaras* Law Journal, 1977, 8.

It is apparent then, some groups are more powerful than others and from significant lack of pollution laws on specific environmental dangers, from the weakness of existing legislation and its enforcement, we suggest that state of legislation not only favours but has been strongly influenced by; powerful, industrial and governmental forces whose economic base is of financial and industrial capital.

(*iii*) *Penal Sanction* : There were certain minor offences which did not contaminate the environment or were not directly responsible for it. For such offence the punishment was imprisonment for a term which may extend to 3 months or with fine which may extend to Rs. 1000 or both.[42]

The major penalty or punishment may extend to six year and fine upto Rs. 1000 or both.[43]

The Environment (Protection) Act, 1986 enhanced the penalty. For the non-compliance of provisions of Act, the punishment is imprisonment extending to five years or fine exceeding Rupees one lakhs or both.[44] There is also provision for enhanced penalty of fine of Rs. 5000 per day in case of offence continuing.[45]

The Environment (Protection) Act, 1986, successfully lifts the veil of corporate personality and it provides that if the offence is committed by a company then every person who was directly in charge of the business of the company, at the time of offence, as well the company shall be proceeded against and punished accordingly.[46] Even government departments are not exempted from the operation of the Act.[47] Emission of the air pollutant in excess of standard prescribed is also penalised.[48]

Other Sanctions : Apart from the penal action, other means are also available to control pollution. They include :

---

42. Section 42 of Water Act, 1974, Sections 33, 37, 38, 39 of the Air Act, 1981.
43. See Section 43 read with Section 24 ; Section 44 read with Sections 25 and 26 of Water Act, 1974.
44. Environment (Protection) Act, 1986 Section 15(1).
45. *Id* Section 15(2) and also See Section 33-A.
46. Environment (Protection) Act, 1986, Section 16(1).
47. *Id* Section 17.
48. Air Act, 1981, Section 22.

(*i*) Firstly, publication of name of offender,[49]

(*ii*) Second type of sanction is provided in Section 5 of Environment Act, 1986. It empowers the Central Government to direct

(a) the closure, prohibition or regulation to any industry operation or process ;

(*b*) stoppage or regulation of supply of electricity or water or any service.

We may sum up by saying that the general legislations and environment legislations are sufficient to deal with any eventuality or industrial disaster, what is lacking is, proper implementation and above all the 'lack of interest' on the part of law enforcing authorities to enforce the aforesaid provisions.

3. Law Provisions

Water (Prevention and Control of Pollution) Act, 1974, and Air (Prevention and Control of Pollution) Act, 1981[50] and Environment (Protection) Act, 1986 contain provisions restricting industries from polluting the environment.

Procedure regarding hazardous substance has been laid in the Act. It also takes care of excess pollutant safeguards in compliance with hazardous substance and other related matter.[51]

Environment (Protection) Act, 1986 excludes the jurisdiction of the court against government or any other authority for compliance of provision of the Act.

The Act, 1986 also restricts the access of public to the court against polluting industries, as it provides that no court shall take cognizance of an offence unless the Central Government complains or sixty days notice has been given.[52]

The enactment of Environment (Protection) Act, 1986 has also enlarged the role of Central and State Boards. Effluent and emission standards have been notified under these Acts in respect of 31 specific industries - 15 for air and 16 for water.

---

49. Water Act, 1974, Section 46.
50. The Water Act, 1974, Section 25 and Air Act, 1981, Section 21.
51. Environment (Protection) Act, 1986, Section 25.
52. *Id* Section 25.

It may be pointed out that statutory requirement of observing the standard prescribed for the industrial pollutants does not make any distinction between industries. The suggestion of Dr. Manmohan Singh, the present Finance Minister may be worth consideration. He has advocated that the standard should be prescribed industry-wise,[53] this will, no doubt, involve a lot of home work for which government will give an excuse but patient-wise treatment would bring in good result rather than one treatment for common disease.

For industries where specific standard has not been notified general standards for discharge of effluents has been notified gradually, in industries specific standards are planned to be notified by other industries as well. Under the National Ambient Air Quality, Monitoring Stations are presently in the operation throughout the country. The government is encouraging setting up of common effluent treatment plant for medium and small scale industries.

Apart from the above, control, the 1987 amendment to the Air Pollution Act in Section 22-A authorises the board to make an application to a court not below that of Metropolitan Magistrate or Judicial Magistrate of first class for restraining any person from emitting air pollutants. Thus the new provisions would enable the authorities to move the court instead of waiting for person aggrieved or interested in protecting the environment.

Establishment of Environmental Laboratories

Water (Prevention and Control of Pollution) Act[54], 1974, Air (Prevention and Control of Pollution) Act, 1981[55] and Environmental Protection Act, 1986[56] have elaborate provisions for establishing environmental laboratories such laboratories will analyse or test sample of air, water, soil or other substance after the sample is submitted to it to ascertain the degree of pollution.

83 laboratories located in various parts of the country have so far been recognised as environmental laboratories.

---

53. Dr. Manmohan Singh, Environment and the New Economic Policies, 36 *Yojana*, 1992, 8.
54. Water Act, 1974, Sections 51 and 52.
55. Air Act, 1981, Section 17(2).
56. Environment (Protection) Act, 1986, Section 12.

It may be pointed out that the Ministry of Environment and Forest visualise the considerable adverse impact on environment which was causing a concern for the very survival of living beings. In view of this the Ministry issued notification on 29th January 1992 imposing rigid conditions on the hazardous industries. Now it was mandatory for any project mentioned in Schedule I and II to take environmental clearance from the Central and State Governments respectively. These schedules enumerated 24 projects each. The owner of the hazardous industries was required to submit an environmental impact assessment report, and environmental management plan for the clearance of his project. In the wave length of rigid control on the hazardous industries the Environment (Protection) Second Amendment Rules, 1992 imposes a further condition of submission of environment audit report by the owner every year on or before 15th May.

4. Response of Judiciary

Indian Supreme Court has tried to check this 'unholy alliance' of development through hazardous industries through its trend setting decisions to protect and preserve the environment. Responding to this wave length in the writ petition of *M.C. Mehta*,[57] the Supreme Court initially ordered to close down the industries located along the Ganga but it was only later on that the court directed them to take necessary steps to check pollution by December 31, 1991 or face closure. The Supreme Court order delivered on December 10, 1991, applies to all 1500 industries dealing in pharmaceuticals, petrochemicals and sugar products along the river Ganga from the Uttar Pradesh dealing down to the State of West Bengal. These industries both in private and public sector dumped the toxic effluents into the sacred river. The surprising aspect was that the State pollution control boards continued to watch them helplessly and even the much publicised Ganga Action Plan did not yield the desired results.

It is also interesting to note that 28 breweries in Uttar Pradesh had earlier been told by Supreme Court to set up treatment plant by 31st December, 1991 to check chemical pollution.

---

57. Reported in *The Times of India*, December 22, 1991.

The Indian Environmental Minister Kamalnath had announced[58] in Indian Parliament on December 1, 1991 that his ministry was identifying 'heavily polluted areas' and a specialised plan was underway to control pollution in such areas. He cited the example of Talcher, Brajnagar and Behrampur in Orissa. He further added that the Ministry of Environment had identified 17 heavily polluting sector including chemical and leather industries. They had been given notice to meet the prescribed standard by December 31, 1991 and they were warned that stringent action would be taken against them. It is pity that in spite of all those 'lofty ideas' to prevent the environment from obnoxious and toxic emissions of hazardous industries, the honourable Minister seems to succumb to the pressure of vested industries and announced the postponement of the new policy standard till 1992. Thus at one stroke, executive put in abeyance the 'achievements' of Judiciary to protect and preserve the environment.

A five judge constitution bench of Supreme Court in *Union Carbide* case[59] on 3 October, 1991 upheld the £470 million settlement in the Bhopal gas tragedy. The bench also directed the prosecution of union carbide officials responsible for world's worst industrial disaster.

The court which passed several directions on properly administering Rs 1200 crore settlement fund for the benefit of victims, also directed the Central Government to "make good deficiency if the settlement fund was found to be insufficient."

In this case, Supreme Court issued several directions. Some directions were impracticable and could not be carried out in practice.

The Supreme Court's ruling that there will not be any upper limit for insurance liability which is expected to cover at least population of one lakh falls in this category. It has also directed the government to bring under insurance cover those who are at present asymptomatic and have not filed any claim but might develop symptom in future. The scheme is also expected to cover children who are yet to born and may manifest cogenital MIC (Methyl isocynate) related diseases.

---

58. See *The Times of India*, December 2, 1991.
59. AIR 1992 SC 248.

It appears that Supreme Court has not taken into account, the practical difficulty in implementing the above direction.

A similar Act passed by Parliament in 1990 making it mandatory for all industries handling "hazardous substance" to go in for unlimited public insurance policy had not been implemented for the same reason.

The Act which was brain child of the then Minister of State for Environment, Mrs. Maneka Gandhi was introduced with Bhopal tragedy in mind and required that hazardous industries should pay compensation to the victim of any such disaster even if there was no fault of industry concerned.

The law seems to be impracticable because the insurance companies are in no position to under write any risk without an upper limit.

We may sum up by saying that the above mentioned factors' have proved to be stumbling block to those seeking to achieve aradical change in the regulation of hazardous industries.

5. Position in M.C. Mehta

(*a*) *The Sriram Industries*

In *M.C. Mehta* case, a public interest litigation was filed under article 32 seeking for closure of Sriram food and fertilizer industries engaged in manufacture of hazardous products. In this context, Government of India had given an assurance in Indian Parliament that government was aware of the 'hazardous nature' of various plants of Sriram and had taken necessary steps for securing observance of safety standard.

Sriram Food and Fertilizer Industries had several units engaged in manufacturing and processing hazardous and lethal chemical and also gases including chlorine, oleum, superchlorine, phosphate at their establishment in Delhi.

'Hazard' is defined as the potential of a chemical to cause an adverse health effect as a result of sufficient exposure. Hazard assessment involves the determination of nature and extent of adverse health or environmental effects associated with exposure to the pollutant in question.[60] Since Sriram was engaged in processing and manufacturing of hazardous product like oleum, chlorine, these hazardous products were capable of adversely affecting health of people living in the vicinity of the plant or could cause extensive environmental degradation. There was leakage of oleum gas on 4th and 6th December, 1985, from caustic chlorine plant of Sriram, thus endangering the life and health of workmen and general public living in the vicinity of the plant.

It was claimed by the petitioner that 'oleum leakage' had affected a large number of persons both amongst the workmen and the public living in the vicinity and caused death of an advocate practicing in the Tis Hazari Court.

Since chemical (including chlorine, oleum and other products manufactured by Sriram) and fertilizer found a place (as items 18 and 19 respectively) in first schedule of Industrial (Development and Regulation Act, 1951, Sriram was confirmed to be an hazardous industries.

The above facts supports the view that Sriram was carrying an industry which was 'hazardous in nature !

'Sriram' had obtained 'consent order' as per provision of Section 25 of Water (Prevention and Control of Pollution) Act, 1974. Persuant to the consent order Sriram installed affluent treatment plant in vanaspati, bleaching powder, superphosphate and active earth units with a view to complying with limiting standards stipulated by the Central Board in consent order.

The Central Board was not satisfied with measures adopted by Sriram for waste water effluent. Sriram installed a 'plant' according to recommendation of the Central Board. However, the said plant failed to give desired result. Sriram installed at their terminal outlet another plant based on air flotation technology, Central Board expressed satisfaction about this plant. Sriram had also complied with all the recommendation of various committees appointed by the court or otherwise, necessary for protection of health of people and preservation of environment.

60. Kannan, Krishnan, *Fundamentals of Environmental Pollution*, 1991, 310.

Sriram was registered under the Industries (Development and Regulation) Act, 1951, so its activities were subject to extensive and detail control and supervision by the government. Under the Act, a license was necessary for establishment of a new Industrial undertaking or carrying on any scheduled industry included in the first schedule of the Act. The government could prevent over concentration in a particular region or over investment in a particular industry. The Act also empowered the government to specify the capacity in the 'license' to prevent over development of industry beyond targeted capacity. Similarly Sections 18A, 18AA and 18G of Industrial (Development and Regulation) Act, 1951 had detailed provisions to control and regulate hazardous industries like 'Sriram'.

Since Sriram was confirmed to be hazardous industries and the government could manage and control its operation under the Act. The industries also attracted the industrial policy resolution, 1948 as Sriram 'Food and Fertilizer Industries' was carrying on an industry relating to chemicals and fertilizer, it was subject to control and supervision of Central Government. Sriram was also subject to extensive control under the environmental legislations.

The Central Board issued a notification under Section 19(i) of the Air Act, 1981, notifying certain areas in Union Territory of Delhi as air pollution control area. The plants of Sriram were in the air pollution control area and industries carried on by Sriram also fell within the schedule of industries specified in the Air Act, 1981. Sriram was, therefore, required, to apply for 'consent order' under Section 21 of the Air Act which was accorded by the Central Board on 13th June 1985 subject to condition set out in 'consent order'.

The Delhi Administration activated itself after the leakage of oleum gas and issued order for closure of Sriram Industries. The Inspector of Factories, Delhi also issued order for closure of the 'Sriram Industry' till adequate safety measures were adopted. Thus it is evident that law enforcing agencies reacted after the leakage of oleum gas. The role of Delhi Municipal Corporation is not different from the law enforcing agencies. Municipal authorities had not taken care to clear the chocked sewer at terminal point of said factory in Najafgarh area despite the repeated requests from 'Sriram'. Supreme Court expressed its 'unhappiness' towards this 'inaction and apathy of Delhi Municipal Corporation.

In this whole affair, judiciary has played a commendable role. It had balanced the social need with other competing values like protection and preservation of environment.

Supreme Court appointed a number of high power expert committees to go into depth of the matter, after filing of public interest petition by M.C. Mehta. Subsequently it gave direction to Sriram to follow the various recommendations of high powered committees. It also satisfied itself through report of Dr. Sharma Committee, about proper implementation of its directive by Sriram. After satisfying itself, that all necessary measures were adopted by Sriram, it permitted the Sriram Industry to reopen the caustic chlorine plant with stringent conditions but it modified some of the stringent condition of first order on the request of Sriram Industries.

In the instant case[61] the Supreme Court emphasized the need to evolve a national policy by the government for location of toxic and hazardous industries and it further added that industries to eliminate the risk from operation of industries. In this context, it advised that there should be preferably green belt of 1 to 5 km around such industries.[62]

Thus it may be said that judiciary has played its part well and live up to expectation of all sections of society.

(b) An Appraisal of the Control on Sriram

It is interesting to note that in spite of the concern of the Indian Parliament and also the presence of multiple agencies to combat environmental pollution ; the leakage of oleum gas from Sriram plant did take place. The Central Board which had main responsibility to enforce regulations strictly only activised itself only after the leakage took place. *M.C. Mehta's* case had exposed the inherent weakness of toothless legislations and inactive executives to regulate and control the hazardous industries. However, the credit goes to the Indian judiciary which compelled the 'Sriram Industries' to install modern control equipment based on latest technology to combat the environmental pollution. The *M.C. Mehta* case also shows that there was a lack of coordination in different governmental agencies in regulating the hazardous industries.

---

61. *M.C. Mehta* vs *Union of India* AIR 1987 SC 965.
62. *Id* at 981.

We may sum up by saying that different government agencies were often working for cross purposes without coordinating their activity.

(c) *Impact of Sanction on Sriram*

Sriram 'Food and Fertilizer Industries' had been held guilty of negligence and was convicted for failing to take preventive measures to control environmental pollution in a judgement of far reaching consequences delivered by Metropolitan Magistrate, Delhi on 9th June, 1992. The order which was passed by the Metropolitan Magistrate Z.S. Lohat after 64 hearings spread over six and a half years and the result of such a long awaited verdict was a fine of Rs. 100 prescribed under the 35 year old Delhi Municipal Corporation Act, 1957. Further, this fine was to be paid by the company and four of its employees.

The judgement vividly explains the inadequacy of old laws in twentieth century to control and regulate the hazardous industries. It is also plain truth that neither the company, nor the general manager, or the maintenance engineer or work supervisor would have any impact of the light sentence. It is desirable that Municipal authorities in Delhi and elsewhere should dust off their old regulations and introduce much stiffer fines for the negligent industries which even go to the extent of causing death. It is also necessary that our law makers should consider the much needed reform and introduce the concept of unified approach to deal with such activities. The hazardous pollutors must pay heavy cost.

We may sum up the whole situation by quoting the Editorial observation of the Indian Express :

> The Sriram case is a reminder of how much still needs to be done to bring civic laws and regulations in line with sound environmental principle. Apart from improving land use and zoning regulations, monitoring commercial enterprises, the transport industry and thousands of small units in the unorganised sector, Municipalities need to look in to their own practices which themselves are hazard to health and life.[63]

---

63. 'A Ludicrous Sentence' an Editorial in the Indian Express, 13th June 1992.

(*ii*) The Aftermath of Mehta

A prominent social scientist says that first stage in the law making process comes when some person or group sees a set of objective conditions as problematic, posing a danger or containing the seeds of future difficulty.[64]

India is on the threshhold of such difficulties. Recent liberalisation of economy has *opend* the flood gates for the foreign companies including multinationals to establish a number of industries including hazardous industries. Such industries will need large scale inventories of hazardous material and intermediates which will need to be handled at one place ; thus increasing the risk manyfold. Perceiving these risks, Indian legislature propose to amend[65] Environment (Protection) Act, 1986 to face 'new challenges' arising out of rapid pace of industrialisation. Five major amendments are being proposed in the Environment Act, 1986 to meet the felt necessities of the present time.

Section 19 of the Environment Act, 1986 required that no court should take cognizance of any offence under this Act except on a complaint made by any person who had given notice of not less than sixty days. The first amendment seeks to reduce the period of notice from sixty to thirty days. Thus the period of 30 days would make it much easier for an individual complainant to seek early redress. The second amendment seeks to remove *anamolus* provision[66] related to the application of the Environment Act. Now a case involving hazardous substance will attract the provision of 'this Act' and not any other under which it might fall, as was previously the case. The third amendment deals with rules for management of hazardous and toxic waste, thus emphasizing the safety requirement in factory. The next amendment deals with rules of manufacture, import and storage of toxic substance. And the last amendment deals with rules or the transportation of hazardous and toxic substance.

Section 3 of the Environment (Protection) Act, 1986 provided for the establishment of an environment protection authority but

---

64. Becker, H.S. *Outsiders : Studies in the Sociology of Deviance*, 1963.
65. See *The Time of India*, 4 December, 1991
66. Section 24(2) of Environment (Protection) Act, 1986.

since six years have passed no such authority has seen the light of the day. This executive lapses has been defended by the government by arguing that the existing frame work would better implement the law.[67]

There is a strong case for constituting the "Environment protection authority' as originally envisaged under the Act. Such authority will adopt an uniform approach to monitor and assess the impact of pollution on particular area before granting permission to establish individual project. At time such authority could also order for closure of defaulting hazardous industries. It is suggested that the authority should be composed of eminent persons in scientific and technical legal field and should have power and machinery to investigate the complain against hazardous industries.[68]

Another significant development[69] is the Water (Prevention and Control of Pollution) Act, Cess (Amendment) Bill 1991 which contains two tier cess structure with a view to providing a built in incentive for such units which use wasteful technology. Piloted by Kamalnath, Minister of State for Environment and Forest, the new Bill imposes a higher water tariff on those industries whose trade effluents are in excess of the prescribed standards. At the same time, there will be a lower tariff on industries that confirm to the optimum water standards. The main thrust of the Bill is to ensure that industry does not use water indiscriminately. Thus the draft legislation will not only succeed to check the pollution but also fill coffers of state pollution control boards. The historic importance of this amendment is that industries which generate pollutants toxic and non-biodegradable will bear the burnt of tariff increase. The basic philosophy behind this amendment was that polluting industry must compensate community and society.

Another area where old measures against pollution control is likely to fail, is cumulative effect of pollutants. Some of the pollutants may not be detected or may remain initially at the

---

67. See *The Times of India*, 4th December 1991.
68. Environmental Law Experts have recommended the establishment of National Environment Protection Agency, See *Environmental Protection Act an Agenda for implementation* prepared by Upendra Baxi, 1987, 8.
69. See *The Times of India*, 2nd December, 1991.

acceptable level, but because of cumulative effect on absorption by certain organisms their concentration may sometime reach the critical level on prolong exposure or through some of the food cycle.

K.I. Vasu suggests a new legal strategy that requires an interactive understanding of dynamic roles of scientific political and legal tripoid of environmental protection.[70] There is one more dimension of the problem which has been pointed out by Madhavan Pillai[71] that the products of biotechnology have yet to attract the provisions of the Environment Protection Act, 1986. This aspect needs serious consideration.

Justice Krishna Iyer says that pathology of legal importance in overcoming pollution is not so much that we have not enough laws but that most laws bark but don't bite. According to him, the reasons are four fold.[72]

1. Environmental justice is entrusted to bureaucratic machinery which has passion for files, not for people.
2. People's agencies even association of non-official expert cannot directly file a complain and board is bottleneck.
3. The most corporate offender have political and financial clout, so board's bow never beat them.
4. The law is genetically so clumsy that quick action is lawlesness.

We may sum up by saying that executive and bureaucrates should wokeup from their long slumber and recognise the danger posed by 'hazardous industries' otherwise we would have to prepare ourselves to face more 'tragic industrial disaster' than the Bhopal gas tragedy or oleum gas leak resulting in innumerable loss to human lives and ecology.

---

70. K.I. Vasu : Where the law is likely to fail ; *Law and Environment* (1992) Editor Leela Krishnan.
71. See *Law and Environment,* Editor Leela Krishnan, 1992,250.
72. V.R. Krishna Iyer, *Environmental Pollution* and the Law, 1984, 26.

PART VI : CONCLUSION AND RECOMMENDATIONS

Conclusion and Recommendations

Industry, and in particular, the hazardous industry is the first class enemy of the environment. The earth is facing today a great challenge of its age to sustain the environment. Therefore, the present generation must rise from slumber and must make all out efforts to save the nature and its beauty. Unless this is accepted by the industrial world as the goal of excellence, the fast development, more civilization and more power would be meaningless. These very objectives would bring the living earth near the dooms-day. It is a pious obligation of every one of us to make our mother earth happy, wealthy and prosperous.

India has joined the mad rush of development process with the western world. The Indian industries have attained tenth position in the world gross industrial output. In this contribution of Indian industries, the Government of India has further accelerated the developmental process by inviting the foreign and multinational corporation to exploit our mother land. This will bring more sufferings to our *matribhumi* and 'we the people of India'. Today there are seasonal changes, more diseases, the Run of Kutch is extending its territorial boundaries, the rare plants and birds, and the flare and fauna are on the verge of extinction, the green belt are loosing their dimension and what not. Can we not imagine the dimensions of the above results for the future? The necessity of the day is not just one way traffic in the direction of industrial progress but a balanced approach between prosperity and environment. It is time that the government must come out of the politics of economy and seriously and sincerely make efforts to achieve the constitutional mandate to "protect and improve the environment." No slogan but *karma* (action) will make *Bharat mahan*. Let us join hands for *matri-seva* in this regard.

The *M.C. Mehta* case is a mile stone in not only the history of environmental justice but also the environment in general. The court exposed the output of the Indian industries. It is an eye opener to the industrialists, the government and all those who love the beauty of environment. *Mehta* is an illustration as a how best capitalist can exploit the nature with minimum cost and maximum profit. The multiprocess chemical establishment was situated under one roof. It was surrounded by thickly populated area of the capital of India, Delhi. In this localised industries, a large number of workers were employed under the big establishment of the Śriram complex. It was surprising that all these activities were carried on before the eyes of Indian Parliament, the Central and State authorities, the pollution controlling machineries, the municipal corporation and last but not the least the Delhi based social service environmentalists. It was a single man's battle in the present case where M.C. Mehta dared to drag the big industrialist to the doorsteps of the highest court of the country and invited the judiciary to pass the order preventing the capitalist from polluting the environment.

The above conclusions make it necessary to think for future directions. In no case an industrial complex be allowed to be established in the residential zones. If such a complex exists in such areas, it should be ordered to be shifted to the industrial belt within a specified period. However, the government, wherever necessary, must give an helping hand in this regard. All the agencies, who have the constitutional obligation towards the protection of environment, must continuously evaluate the position of existing environment. This can be done by auditing the protection and pollution control measures. In this regard it is further suggested that an environmental ombudsman may be constituted by reviving the dormant provision contained in section 3 of Environment (Protection) Act, 1986. A well organized occupational health set-up is a must for every hazardous industries. Area monitoring (measurement of concentration of pollutant in a given area), personal monitoring (measurement of pollutant in the immediate vicinity of an individual worker) and biological monitoring (estimation of concentration of pollutant in blood, urine, etc., of industrial worker) should be made compulsory by passing the appropriate rules for the hazardous industries. Highly toxic substances are being handled without proper care and precaution enhancing the danger to human lives and environment.

It is suggested that specific monitors should be installed to continuously measure the levels of toxic pollutants so that the general public and workers can be alerted in case of any drastic change in the pollutant monitored. It is also recommended that if a pollutant is spotted to be potentially hazardous based on toxicological assessment, the pollutant can either be isolated at the point of release or its production and converted for other useful purposes. Moreover, the release of highly toxic chemical in environment should be blocked through various scientific methods available and can be destroyed by specific process in restricted area.

An uniform national policy for the location of hazardous and toxic industries should be immediately formulated by the Central Government. Human settlement should not be allowed to develop upto 5 km from such hazardous industries. Such industrial zone should be declared as prohibited zone for general public habitat. There should be preferably a green belt of 1 to 5 km around such industries. The green belt should include trees of pine, deodar far and other medicinal plants as they have capacity to absorb toxic gases.

It is the function of the State to see that the developmental process subserves the common good and at the same time does not adversely affect the environment. In this function of the State comes the industrial development which is carried on by the State owned industries, State financial and controlled industries and private industries. When any industry steps into the shoes of this governmental function, then it will attract article 12 which enlists the authorities which come within the definition of 'state' for the purposes of Part III of the Constitution of India. It was unfortunate that the Supreme Court left the question open whether Sriram was 'the state'. Looking to the functions, shape, size and output, Sriram was the state within article 12 for the purpose of protection of the fundamental rights. And thus the provisions of article 32 were available in the instant case. Such sanction would keep the capitalist alive to the environmental problem. This analogy stretched further would also attract their constitutional obligation as 'the state' to protect the environment.

Sriram had the Indian share holders. This means that it was a collective efforts of the citizen of India on piercing the corporate veil, Sriram was not different from citizen under article 51A. The Supreme Court of India in the *Benett Coleman*[1] case approved this approach. And if this reasoning is accepted as valid then Sriram was under a constitutional duty to protect and improve the environment. A point which was completely ignored in the *Mehta* case.

The senior advocate of Supreme Court of India and leading environmental lawyer, M.C. Mehta, in the present case deserves all appreciations for performing his fundamental duty towards the protection of environment by successfully instituting the present litigation. It is time that other lawyers must make the constitutional duty a reality.

The constitutional duty to protect and improve the environment would remain incomplete unless a corresponding right also gets a specific place in the Constitution. The judicial innovation in this area needs a constitutional stamp. The citizen's duty may be correlated with the citizen's freedom to live in a clean environment subject to the reasonable restriction in the interests of the general public. This must find a place in article 19 of the Constitution of India. This will maintain a balance between the right and duty and also the environment and development.

The environmental education, which has been given a place in the environmental legislations, the environmental policy and planning, has yet to come out of the paper work and to percolate down to the masses. The illeteracy in this area has allowed the environmental protection and improvement plans to move with a tortoise speed. The literacy has to spread not only in urban but down to gross root level as well.

The Ministry at the Central and State levels, responsible for the education, must create a separate cell for environmental education and it must facilitate the right to environmental education to every Indian citizen. This will make active the dormant obligation of State in article 41. The government must make environmental education a compulsory subject in all the educational institutions. The social activists group and the lovers of environment must come

---

1. AIR 1973 SC 106.

forward to eradicate the environmental illiteracy. Once the masses are aware, no force can stop them from making India more beautiful. In this respect *Mehta* has educated the judges, lawyers, both the parties and all those who has and will read the judgement. *Mehta* is a milestone in the environmental education.

Coming to the administration of environmental justice, the Indian judiciary, even though well equipped to handle the environment cases, feels handicapped in balancing the technological advancement and the pollution of environment. The *Mehta* case is an eye opener in this regard where judges time and again consulted and were assisted by one after another expert committees. In spite of this assistance, the court remained still unsatisfied and unconvinced with the recommendation of the leading experts in the field of environment. If the court works under this psychic it may not successfully deliver environmental justice; moreover, the modern technology is fast developing and fast changing and this requires continuous know how. The Indian judges, loaded with burden of enormous pending cases of different varieties, cannot keep themselves abreast with neo-developments. It is suggested that the judges should be given some exposure to these developments from time to time by organizing seminars, lectures of eminent person in the field of environment. A refresher course may be organized for the judges to make them familiar with fundamentals of environmental pollution engineering and management. This case study has emphasized that it is prerequisite for the judges, handling the cases in this area, to have scientific and technical background before deciding the dispute relating to environmental pollution. The suggestion of Supreme Court for establishment of ecological science research group should be immediately implemented. The said centre will assist the court and various authorities in handling scientific and technical environmental issues.

Another factor affecting the administration of environmental justice is the economic philosophy of individual judges. It implies that decision of Supreme Court will lack uniformity. The outcome of a case will vary from case to case depending upon economic philosophy of the individual judges constituting the bench. It is clear from the study of case law of environmental pollution that some judges supported the extreme view in favour of development alone and some other judges tilted the balance exlusively on the side of environment. This bias approach of the judges to suit their own philosophy will be great hinderance in the administration of environmental justice. Therefore, it is suggested that the judges must keep their mind open and must not be influenced with one side or the other. It is necessary that environmental justice must not be seen to be delivered but it must properly balance the interests. In environmental justice, the judges must remember that it is the coordinated and cooperative approach in the development and environment which will sustain the earth.

The view of the *Mehta* experience, it is suggested that the environmental law court must be constituted. The court will have a gross-root level State appellate body and finally a central environmental law court. Initially it shall consist of at least three members one of which shall be a judge of the appropriate level having some scientific and technical temper and rest of the members shall be eminent persons in the field of environment. Moreover, in case of difficulty in handling some exceptional cases, the court may be allowed to coopt expert member in the required field.

It is surprising that highest legislative body in India is moving with a *tortoise* speed. The Environment Court Bill, 1990 has been pending in Parliament for 2 years and it has yet to see the light of the day.

All those lovers of environment including the member of Parliament and public must come forward and awaken the highest legislature of India to transform the Bill into an Act so that environmental court gets a place in the existing judicial organization. India has taken already lead in incorporating fundamental right, fundamental duty and State obligation with respect to environment, now it must take lead in establishing the environmental law courts.

In order to approach the court in the environmental pollution cases, the normal rigid technicalities shall be given up otherwise it will frustrate the very object of speedy environmental justice. Therefore, it is suggested that public interest litigation must be allowed from the gross-root to the central level. In the public interest litigation the court must encourage the group action.

The social environmentalist may at a time need legal aid. In such litigation of general interest of public, the government as well as the court must make available free legal aid in the environmental litigation. The lawyers must devote some time to such litigation free of cost so that they can also contribute not only to their own propriety but also for the environment which is also affecting them. The lawyers like M.C. Mehta deserves recognition for the services rendered in the field of environmental justice. Such a recognition will encourage other lawyers to help in the administration of environmental justice. The court at the same time must scrutinise carefully the public interest petition so that no unwanted element touch the balance of environmental justice for his personal interest.

It is normally seen that function of the court comes to an end when the court delivers the judgement. Thereafter the court normally does not see whether their order is implemented in the right spirit or whether the benefit of the judgement reaches down to the beneficiary. In the environmental pollution cases, it is necessary that there should be continuous evaluation and auditing regarding the implementation of the judgement. The court cannot shut its eyes once the judgement is delivered but it has to see the effect on the beneficiaries and the environment. Therefore, it is suggested that after judgement is delivered in environmental litigation, the court must constitute a committee to appraise the court of the implementation of its judgement in environmental pollution.

M.C. Mehta will be remembered in the history of administration of environmental justice for changing the dimension of a well settled principle of law laid down in *Ryland* vs *Fletcher* which dominated the field for a long time. In case of environmental pollution, the polluting agencies have in many cases, given scot free in view of exceptions provided in the environmental law. This has resulted in allowing the capitalist to earn maximum profit at the cost of environment. All those who are participating in the evil of degrading the environment must compensate the loss to the environment. Their liability should be absolute because the very nature of a hazardous industry denotes adverse consequences on the environment. The absolute liability will compensate to the certain extent to all those who suffer from environmental pollution and it will make the capitalist alert to see that the existing environment is not badly polluted and that precautionary measures in the development process are taken in the real spirit. This principle will bring in the knot the persons who are involved in the development supervision and control processes of the development. Thus the doctrine of absolute liability will percolate down to even the officers-in-charge of the polluting units. It is further suggested that the doctrine may also be applicable in case of governmental agencies who abate or help the industries in polluting the environment.

The compensatory remedies is one of the solution to off-set the losses caused by degradation of the environment. In calculating the compensation, the court must look to the impact of environmental pollution as a whole and also include all those who did or could not reach to the court.

The *M.C. Mehta* justice has yet to reach to all the oleum leak sufferers. It is suggested that the court must adopt speedy process to see that the compensation really compensate those who have suffered and who will suffer because of environmental pollution. The compensation amount will have no meaning when it reaches to the person who is already dead. The *M.C. Mehta* and the *Bhopal mass* disaster are the two burning examples of the delayed compensation. A delayed compensation is in fact no compensation in the real sense. The court must treat the *M.C. Mehta* case a starting point in administering compensation. It may be pointed out that compensation should not be treated as a complete solution in the field of environmental pollution. It must be treated as an interim arrangement and continuous and provision should be made for continuous evaluation of losses occurring from time to time. This is all the more necessary because pollution affects not only the existing component but also the future environment.

Now coming to the first class enemy, the hazardous industries; in many countries they have been black listed. It is time that the Government of India must identify the impact of hazardous industries and black list those who pollutes most. The Constitution of India has distributed 'industry', Industrialization and related matters in lists I, II and III. This approach will allow decentralisation in handling the industrial pollution problems. The individual approaches of various States will create difficulty in handling the industrial hazard. The developed State will not interfere with the development process and the under-develop State will concentrate more and more in the development. In view of the confused and diffused situation, it is suggested that hazardous industries, wherever they may be situated, must be brought under the control of Parliament so that an uniform approach may be adopted in handling the pollution problems.

The legislative *Padyatra* of the environmental law required a special treatment with the hazardous industries. These industries should be clearly identified in the legislation. The existing Central and State Pollution Boards have yet to make any important contribution in handling the hazardous industries. A hazardous industrial commission must be set up to help the pollution control agencies to maintain and control the environmental pollution. This commission will submit an annual report to the Central Pollution Board and also table its report in Parliament. The existing consent provision should be done away with in case of hazardous industries.

It is also suggested that the Chairman of Central Pollution Board or State Pollution Board should not be political person. A distinguished scientist or persons having scientific and technical background should be appointed as Chairman of the Pollution Control Board.

It is the need of the hour to evolve appropriate and innovative economic instruments to combat the menance of environmental pollution. Such economic instruments maybe in the form pollution fees which may be imposed on the defaulting hazardous industries. Financial constraint is main factor which hinders implementation of pollution control measures. Therefore, it is suggested that subsidies may be provided for the cost effective and efficient technologies. These subsidies may be in the form of low interest loans, tax incentive, etc.

It is gratifying to note that IDBI and ICICI are channeling low interest fund from World Bank to help in setting up the common and individual treatment plants and development of cleaner technologies. The existing pollution cess law have resulted in small coins in the big coffers of the State. The cess law requires an amendment with respect to hazardous industries.

The Environmental Protection Act, 1986 incorporated some of the experience of the Bhopal mass disaster but it should be amended to give a special place to the hazardous industries with special sanction. The government must carefully examine the question of localization of hazardous industries and it is only after detailed deliberations inside and outside Parliament that such industrial process be given permission to start. The government must continuously evaluate the environmental impact. The environment impact assessment study should be made compulsory for all high polluting industries. The transportation of hazardous substance is another item which needs careful examination so as to amend the existing environmental laws. The Insurance Companies of India have yet to play their social role. The hazardous industrial process in its wider sense must attract the attention of the insurance organization. This will give such insurance which will give immediate relief to those who have suffered because of the hazardous process. Such insurance should be made compulsory in all the hazardous industries.

Recently, the Bureau of Indian Standard has started to award Ecomark to environmentally friendly products. It intended to raise environmental consciousness of the people as consumer. The Ministry of Environment and Forest has released a list of items for which this ecomark has been made compulsory. Similarly, it is suggested that the hazardous substances should be given a distinct identity so that people may know that they are pollution prone substances.

The outcome in *Mehta* vividly explains the inadequacy of the old environmental laws in controlling and regulating hazardous industries for the twentieth century. Therefore, it is suggested that legislatures including the municipal authorities should dust off their old laws and regulations and introduce much stiffer fines for the negligent industries.

We may sum up, the Indian legislature and executive should recognise the latest developments in the comparative environmental law and adopt measures to upgrade existing laws and regulation to meet the 'new challenges' of twentieth century. The judiciary must also rise to this occasion. In order to make their hands more effective, it is necessary that a strong public opinion must be developed to put a check on the hazardous industries. If we can build up a strong public current against the environmental pollution, we will be fulfilling our pious obligation not only to our motherland but also humanity at large.

Bibliography

I. BOOKS

Agarwal, S.L., *Legal Control of Environmental Pollution*, 1980.

Agrawala, S.K., *Public Interest Litigation: A Critique*, 1985

Baxi, Upendra, *Inconvenient Forum and Convenient Catastrophe: The Bhopal Case*, 1986

Baxi, Upendra, *Environmental Protection Act, An Agenda for Implementation*, 1987.

Baxi, Upendra and Paul, Thomas, *Mass Disaster and the Multinational Liability: The Bhopal Case*, 1986

Baxi, Upendra and Dhanda, Amit, *Valiant Victims and the Lethal Litigation : The Bhopal Case*, 1990.

Backer, H.S., *Outsider : Studies in Sociology of deviance*, 1963.

D'arryl D, Monte, *Temples or Tombs*, 1986.

Diwan, Paras, *Environmental Protection*, 1987.

Gayatri Singh and Madhukar Singh, *Environmental Activities*, 1988.

Harrison, Paul, *The Third World Tomorrow*

Jain, M.P., *Indian Constitutional Law*, 1987

Krishna Iyer, V.R, *Our Courts on Trial*, 1987.

Krishna Iyer, V.R, *Law v Justice*, 1981.

Krishna Iyer, V.R, *Environmental Pollution and The Law*, 1984.

Krishna Iyer, V.R, *Environmental Protection and Legal Defence*, 1992.

Kannan, Krishnan, *Fundamental of Environmental Pollution*, 1991.

Leela, Krishnan, P., *Law and Environment*, 1992.

Lave, L., *Quantitative Risk Assessment in Regulation*, 1982.

Musharraf Ali, S., *Legal Aspect of Environmental Pollution and its Management*, 1992.

Mishra, Arun Kumar, *The Indian Water Pollution Law : Retrospect and Prospect*. An unpublished LL.M. Dissertation of Banaras Hindu University, 1987.

Pearce, *Blue Print for Green Economy*, 1989.

Pandey, Rammohan, "*Prajna*" Magazine of Banaras Hindu University, 1989.

Regers, W.V.H., *Winfield and Jolowicz on Torts*, 1989.

Sapru, R.K. and Bhardwaj, Shyama, *The New Environmental Age*, 1990.

Sathe, S.P., *The Right to Know*, 1991.

Taylor, S., *Making Bureaucracies Think : The Environmental Impact Statements Administrative Reform*, 1984.

II. ARTICLES

Agrawala, S.K., Environmental Pollution : Issues for Consideration, *Legal Aspect of Environmental Pollution and its Management*, ed. S. Musharraf Ali, 1992, 14.

Bhardwaj, Shyama, Management of Environmental Pollution in India, *The New Environmental Age*, 1990, 65-81.

Gunnigham, N., *Power, Politics and Environment, 13 Banaras Law Journal*, 1977, 8.

Jariwala, C.M., The Constitution 42nd Amendment and the Environment : *Legal Control of Environmental Pollution*, ed. S.L. Agarwal, 1980, 7.

Jariwala, C.M., The Changing Dimensation of Indian Environmental Law. *Law and Environment*, ed. P. Leela Krishnan, 1992, 13.

Prof. (Kr.) Khan, Gulam Ahmad, Environmental Pollution : Debate and Challenges, *Legal Aspect of Environmental Pollution and its management*, ed. S. Musharraf Ali, 1992, 6.

Leela, Krishnan, P., Public Participation in Environmental Decision Making. *Law and Environment*, 1992, 162.

Mehdi, Ali, Public Liability Insurance Act, 1991 : A Critical Appraisal, *AIR 1992, Journal 36.*

Muchlinski, P.T., The Right to Development and Industrialization of Less Developed Countries : The Case for Major Industrial Accidents Involving Foreign Owned Corporation, Human Right Unit Occasional Paper Common Wealth Secretariat, London, 8.

Muchlinski, P.T., The Bhopal Case : Controlling Ultra Hazardous Industrial Activity Undertaken by Foreign Investor 50 *Modern Law Review*, 645.

Newark, Non-natural users and Ryland v Fletcher 24 *Modern Law Review*, 1961, 557, 561.

Purduo, Michall, Integrated Pollution Control in Environment Protection Act, 1990 : A Coming of Age of Environmental law, 1991, 54 *Modern Law Review*, 534, 539.

Robinson, N.A., SEGURA'S Siblings Precedents From Little NEPA in The Sister States 46 *Albany Law Review*, 1982, 1155.

Robinson, N.A., Marshalling Environmental Law to Resolve The Himalayan Ganges Problem 13 *Delhi Law Review*, 1991. 1, 24

Sandes, Philip, European Community Environmental Law : Legislation. The European Court of Justice and Common Interest Group 53, *Modern Law Review*, 1990, 685.

Sangal, P.S., Some Recent Development in the Field of Environment Prospect for the Future, 13 *Delhi Law Review*, 1991, 49-71.

Sarkar, N., Industry Alone not to Blame *The Hindu Survey of the Environment*, 1992, 137.

Singh, Shekhar, The Global Environmental Debate, *Indian Journal of Public Administration*. Vol. XXV, July-Sept. 1989, No. 3, pp. 363, 371.

Singh S.N., The Public Liability Insurance Act, 1991 : Scope For Making Provision More Effective, 5 *Corporate Law Adviser*, May 1991, 234.

Singh, Dr. Man Mohan, Environment And The New Economic Policies 36, *Yojana*, 1992, 8.

Tennet, Vijay Chandra, Environmental Court : Need of the Hour, 1 *Supreme Court Journal*, 1991, 36.

Vasu, K.I., Cumulative Pollutants : Where The Law is Likely to Fail, *Law and Environment*, 1992, 78, 85.

Wali and Burgess, The Interface Ecology And Law : Science, The Legal Obligation And Public Policy, *Syracuse Journal of International Law And Commerce*, 1985, 12, 221, Syracuse, N.Y.

Winfield, The Myth of Absolute Liability 42 *L.QR* 34, 51.

III. REPORTS

Annual Report 1988-89 (of Central Pollution Board).

Environmental Quality, 1985, U.S.A. 16th *Annual Report.*

The State of India's Environment 1982, *A Citizen's Report* and State of India's Environment 1984-85. *The Second Citizen Report* The State of India's Environment : Thee Third Citizen Report, 1991. (Prep. by The Centre For Science and Environment).

IV. NEWSPAPERS

The Times of India

The Indian Express

The Hindustan Times

V. MAGAZINES

Prajna (Magazine of Banaras Hindu University)

The Lawyer

Yojana

The Hindu Survey of The Environment, 1992

Corporate Law Adviser

Index